Scentwork for Horses

Scentwork for Horses

Rachaël Draaisma

CRC Press is an imprint of the
Taylor & Francis Group, an **informa** business

First edition published 2021
by CRC Press
6000 Broken Sound Parkway NW, Suite 300, Boca Raton, FL 33487-2742
and by CRC Press
2 Park Square, Milton Park, Abingdon, Oxon, OX14 4RN

Library of Congress Cataloging-in-Publication Data

A catalog record for this title has been requested

ISBN: 978-0-367-55298-5 (hbk)
ISBN: 978-0-367-53760-9 (pbk)
ISBN: 978-1-003-09284-1 (ebk)

Typeset in Palatino
by Deanta Global Publishing Services, Chennai, India

Contents

Foreword

Dear readers of this book,

Finally, it is here: the book about mental stimulation for horses. In the 1970s, we started to make use of nose work for dogs, but it was in the very beginning, which caused some funny situations. In the 1980s, we started to use and develop enriched environments for dogs and used them more and more. I also made use of enriched environments for my horses, but only now, more than 30 years later, it is starting to invade the horse world too, still on a small scale. I do hope this book will help people see how important it is for horses' wellbeing and mental health and, maybe above all, see what a wonderful and entertaining world it is for both horses and people.

In the mid-1970s, I owned a horse meant to race, but I managed to let him be out the whole summer in a huge forest area together with some other horses from the stable (which is still the custom for many horses in Norway). I looked after him every day, as the forest is not the safest place to be but better than being trapped in a small paddock the whole summer. He kept healthy, and strolled the forest up and down, enjoying his freedom and getting really good at finding his way around. One day in the heat of the summer, I noticed that the brook had run dry, so the horses had no water. I hurried back to the stable, filled up water cans, and got back as fast as I could. I stopped the car and got out to see where they were, and there, in the middle of a grass slope with only brush around, I saw my horse wander back and forth, nose to the ground. I watched, fascinated, wondering what was going on. He was intense in his search, so I did not want to disturb him. All of a sudden, he stopped and started to dig. The hole became bigger and deeper, until finally a silver stream of fresh water came out of the hole. He had smelled water and dug it out! The horses drank the fresh water, and then continued to graze. I returned to the stable with my water cans, filled with joy about the wonderful senses of horses, the senses that have made them survivors for so many years. The memory still fills me with joy and admiration.

From that moment on, I have let my horses use their senses as much as I can, knowing how important they are, perhaps no longer for survival, but certainly as a part of who they are, naturally, and I want to make them use what they were meant to use, in some way.

My hope is that all horse owners will do that, for the mental health and pleasure of the horses. This book might be the help they need to get going. Thank you, Rachaël, for fulfilling my dream of a mental stimulation book for horses. I never got down to it, but you did!

Your friend Turid Rugaas
President of Pet Dog Trainers of Europe
Founder of TR International Dog Trainer Education
Owner of Hagan Hundeskole
Horse and Dog Behaviour Consultant and Trainer
Speaker - Author
Kristiansund, Norway

Acknowledgements

I consider myself blessed to be able to share the subject that has captured my heart and soul with the world. In addition, I cherish the fact that I can do this side by side with colleagues I also call friends. You do not create a book on your own. It is the sum of all I have learned and experienced, peppered with the knowledge and experience of others who were kind enough to share what they knew with me. This would happen on the phone, in kitchen-table conversations, through Facebook Messenger or email, during vacations, and, of course, during lessons and shared activities. I owe many thanks to Tracie Faa-Thompson, Risë VanFleet, Els Vidts, Ristin Olthof, Suzy Deurinck and Julia Robertson.

A special place in my heart is reserved for Turid Rugaas and Anne Lill Kvam. My work builds on the foundation laid by Turid Rugaas, but Anne Lill Kvam also played a role in this project. I am very grateful to them for their engagement and support. In all things, they are the embodiment of dogs' and horses' right to lives that suit their nature: a vision to which I also subscribe.

Special thanks to Els Vidts whose help was so welcome when conducting the chapter on Brain & Biology.

Special thanks also go to Nanne Boekholt and Julia Robertson, who added terminology and knowledge to what I saw during scentwork. In close cooperation with them, I was able to write the chapter on biomechanics. Nanne Boekholt works as ICREO Animal Osteopath and rehabilitation trainer for horses at De Dierenosteopaat. Julia Robertson is founder and director of Galen Myotherapy, a (global) branch of massage therapy that promotes health and treats muscular pain in dogs through unique massage techniques and exercise management.

Finding studies in which horses' olfactory sense was a main or partial subject took a great deal of time. In this, I received help from Suzy Deurinck. Thank you, Suzy.

My tracking method has led to the creation of a scent bag tailored to this specific purpose. My son, Sybrand Jansen, was responsible for its design, going through a great many fabrics, sizes, and calculations. Many of his prototypes were tested in the real world. When the final prototype arrived, Paul Stam finalised it into the current model. Many thanks, Sybrand and Paul.

I also want to extend warmest thanks to my beta readers: Marian Draaisma, Els Vidts, Nanne Boekholt, Ristin Olthof, Suzy Deurinck, Anne Lill Kvam, Tracie Faa-Thompson, Risë VanFleet, and Julia Robertson. I hold your opinions in high regard, and I am very grateful that you took the time to read my work and share your thoughts.

The many photographs in this book were mostly taken by me or Hans Ruijs. The professional photographers Nikki de Kerf, Méadhbh Ní Dhuinn, and Gijs Timmers also supplied many beautiful images. An identification of their photos appears to be in the About the Author's sectionr.

Many thanks also to the people and horses whose stories I was allowed to tell and whose pictures I was allowed to use. They are Annemarie van der Wal, Esther Janssen, Martine Liefstingh, Nanne Boekholt, Ristin Olthof, Henk Jan Nix, Ymke Michels, Laura Scholte, and Susan Kjaergård. But also thanks to my stablemates and William and Astrid Janssen for their support.

I come from a close-knit family that is very engaged with me and my work. They are my beacon and foundation. Their support, but also the freedom and space they grant me, gives me wings. I especially want to name Marian Draaisma, Bert Peters, Robert Peters, Hans Ruijs, Kitty Draaisma, Ronald and Jasper Draaisma, Taliha Eren and of course my children: Sybrand and Imke Jansen.

Finally, thanks to translator, Sarah Strous, and thanks to my publisher, CRC Press, and, in particular, to Senior Account Manager Alice Oven, whose engagement, enthusiasm and support made our collaboration a pleasure. Thanks!

About the author

Rachaël Draaisma graduated from the Radboud University in Nijmegen, the Netherlands. She has always lived with and had a passion for dogs and horses. In 2002, she decided to make it her profession. Rachaël obtained several diplomas and then started working as a trainer and behavioural consultant, first with dogs and later with horses. In 2013, she completed the TR (Turid Rugaas) International Dog Trainers Education course. Under Turid's supervision, she began studying the calming signals of horses, filming domesticated horses, and analysing the resulting material. The study grew, is ongoing, and led to a complete focus shift from dogs to horses. It also led to the worldwide publication of the book *Language Signs and Calming Signals of Horses* by CRC Press.

Another pillar of Rachaël's working life with horses evolves around equine mental stimulation and scentwork. Rachaël developed an extensive method to do scent tracking with horses, in which she teaches horses to follow a footstep trail, so they can find missing persons or scent bags. Tracking has enormous advantages for a horse's behaviour, brain, and body. But scent tracking with horses is also enriching for humans and can be seen as a new tool in enriching the human–equine relationship. Rachaël develops tracking to a level at which she can use it as a sole (primary) activity for horses. She also uses mental stimulation and scent tracking as a method when working with horses who have behavioural problems.

Rachaël travels throughout Europe and the globe to lecture about the calming signals of horses, equine mental stimulation, and scentwork. She also gives workshops and a series of weekend sessions that combine theory and hands-on work with horses. Many universities and education programmes are interested in her work. China, Denmark, Poland, Belgium, Spain, Slovakia, Sweden, the United Kingdom, and the Netherlands are only a few of the countries she has visited. Rachaël's first book, *Language Signs and Calming Signals of Horses,* is an international best seller and has been translated into several languages.

Credits

The following images are attributed to:

Nikki de Kerf Fotografie
Photographs from the cover
Photos 1.3, 3.36, 6.10, 9.1, 10.2, 14.39, 16.1, 16.2
Photo used for Illustration 7.5

Gijs Timmers Photography
Photos 14.13, 14.43, 15.6, 15.7, 15.8, 15.9, 15.10, 15.11
Photos used for Illustrations 7.1, 7.3, 7.4, 7.6, 7.7, 7.8

Méadhbh Ní Dhuinn, Photographer
Photos used for Illustrations 6.7, 6.8, 6.9

Susan Kjærgård
Photos 4.23, 4.24, 4.25, 4.26 (Michael Ernstsen)

Additional information
Figures 1.2, 1.3, 4.1, and 4.18 are adapted fragments of Toppotijdreis. Source: Kadaster Apeldoorn.

This book was translated from Dutch to English by Sarah Strous, Book Science Editing, Den Haag, the Netherlands.

Introduction

How do you introduce a topic that is so close to your heart, a topic about which you have so much to say that, rather than fail, words gush out, trying to find a place on the page? Do I start with the horse's obvious joy when he gets to do scentwork again, the physical and mental benefits that come with it, its uses for rehabilitation and behaviour therapy, its success at socialising or resocialising horses, or the bit of individuality the horse regains from it? There is so much from which to choose. Exploration and scentwork is so much more than just another game. It is a serious instrument that has an impact on various aspects of a domesticated horse's life, no matter the horse or the life he leads.

But you cannot forget about people in this either. When you are talking about exploration and scentwork for a horse, the handler discovers right along with him. In so doing, you will experience patience, wonderment, surprise, and maybe a little frustration now and then, but hopefully mostly inspiration, admiration, and enjoyment, if you use the elements from the book that suit you, your horse, and your shared lives. Here is to exploration and journeys!

THE PURPOSE OF THE BOOK

The overarching goal of this book is to introduce exploration and scentwork in the life of the horse and establish it as a serious instrument.

The horse lives in a world filled with human stimuli. The degree to which he can handle them influences his wellbeing and the life he leads. The aim of the first part of the book is to give readers tools that they can use to help their horses experience and discover these stimuli. This will empower your horse and allow him to better understand our world, enabling him to handle it better, with a relaxed state of mind, a zest for life, and a proactive attitude. Attention is also paid here to horses who need more care.

The purpose of the second part of the book is putting scentwork with horses on the map as a discipline. Scentwork is beneficial to any horse, no matter the circumstances in which he lives and works.

The many suggestions you will find throughout this book will enable you, the reader, to implement some or all of its elements in your life and work with horses.

Generally speaking, a horse lives a very controlled life in our domesticated world. His life is almost entirely shaped by the plans, desires, and demands of his owner. The goal of this book is to give back a little individuality, choice, and self-determination to the horse, without losing sight of the realities of everyday life.

Letting your horse discover and track requires the handler or rider to take on a different role. You do not steer or control your horse when he is exploring or tracking but rather you are on the sidelines, deciding a fitting setting for your horse. In this, getting to know your horse, finding out his likes and dislikes, what is easy for him and what he finds hard, is an inevitable side effect (and indirect goal) of this book, just like reacquainting yourself with his body features and communicative signals.

STRUCTURE OF THE BOOK

I have chosen to structure the book in the following way:

Part 1 is about exploration. What role does this play in the horse's life? What are the benefits and drawbacks of giving your horse more freedom to discover stimuli and make choices for himself? It begins with establishing your horse's comfort zones. Then, you can use the many exercises to get started, and to do so in such a way that both you and your horse can take optimal advantage of the many good things that come with this. I will also show you how to use exploration as a valuable tool for horses who need extra help and care. For these horses, you will make a special, customised plan. I will give some examples for this, as well as a real-life case I encountered during my work, about which I share general information but do not go into details.

Part 2 deals with scentwork, a discipline that involves a very different kind of exploration. I discuss scent tracking, in which a horse follows a footstep scent trail, and treat search. With these two scentwork exercises, you have a light scent exercise at your fingertips that is fairly easy to set up (treat search), and one that allows you to go into more depth (tracking).

The appendix contains six pages that can be used as practical guides for Parts 1 and 2. For instance, there are lists and graphs that will help you determine your horse's comfort stages with regard to various stimuli as well as the haystack exercise and two checklists you can use for scent tracking. I can imagine that, in this form, the pages are easy to copy and bring along when you set to work with your horse.

In the past, I have also done scent discrimination with horses, for which I taught them to point to a particular scent. This could take place in a larger living area or among other distracting scents; however, I have not delved further into this discipline. I chose scent tracking and treat search instead because they involve a greater sense of autonomy for the horse and a more accommodating role for the handler, even in the early stages of the exercises, which was exactly what I was looking for.

With this array of practical exercises, both for exploration and for scentwork, I hope to offer a varied range that will appeal to many readers and fit in well with all the different ways of working and living with horses.

The practical exercises in Parts 1 and 2 are preceded by theory. That way, you have a framework in which to place the exercises, background information, and a clear idea of why I make certain choices as well as all their upsides and downsides. Also, giving you the theoretical knowledge makes it easier for me to quickly refer to this when I am explaining the practical exercises.

Parts 1 and 2 can be read as separate segments. There is only a small overlap in theoretical knowledge, which makes this possible. However, if you want all the theoretical information, I would read both parts. If you already know the theory or the theoretical stuff is not for you, but you do want to get going on the practical work, then you can also skip the theory and get started straight away with the practical exercises.

HOW WILL THIS BOOK HELP YOU?

This book was written for horse owners, trainers, and behaviour therapists, so that they can work with their own or other people's horses.

The book helps you to guide a horse in exploration and scent tracking. It offers practical tools and exercises, for which pictures are a visual support and their captions provide new information. The Figures 6.1, 6.2, and 6.3 were created especially for this book. They are figurative and meant to give insight into the olfactory process. Certain dimensions, for instance the size of the scent particles, are not true to life.

When learning new skills, you obviously do not want to make mistakes but work as perfectly as you can. That is why I have tried to describe the exercises as extensively as possible. I also offer alternatives and preparatory exercises in case you encounter challenges. Every horse is different, as is every person. We all have knowledge, experiences, and emotions we bring along. It is inevitable that you will come across a situation that is not described in this book. For those cases, I hope that the theory and the practical tips form a framework that helps you find solutions tailored to you and your horse and that benefit you both. Incidentally, this is also one of the charms of exploration and scentwork: As a handler, you are just as busy looking, planning, and directing, thereby also keeping your own thirst for exploration alive and sated.

When you start implementing exploration and tracking in your horse's life, remember that, however effective they are, they cannot replace a visit from the vet, medical specialist, or behaviour therapist. If you or your horse are experiencing discomfort, pain, or high tension, the first step is consulting with specialists.

METHOD

This book is the sum of the knowledge I have gained from living and working with my own dogs and horses, running a dog school, and working as a behaviour therapist first for dogs and now for horses as well as from continuing to take courses, attend lectures and participate in workshops. I received a solid foundation for this from Turid Rugaas's training course (TR International Dog Trainers Education). This course was also the starting point of my ongoing study of the language signs and calming signals of horses, in which Turid is my mentor. This research aligns perfectly with and strengthens my practical work with horses. After all, being able to read a horse is indispensable when you are talking about his mental development, ability to handle stimuli, comfort zones, behaviour, and wellbeing.

The exercises I discuss in Part 1 of this book and the treat search from Part 2 both originate with Turid Rugaas's course, as well as with the work of Anne Lill Kvam, which also shaped and helped me. I personally developed the tracking methodology described in Part 2. Central to this is the practicability of the parts and the whole for horses and people. It is also important to me that the method brings happiness, joy in life, and an eagerness to participate; that it does not cause either the horse or the handler pain or unpleasantness; that it does justice to the horse as an individual; and that it positively influences and strengthens the relationship between horse and handler or rider.

HORSES

The horses and ponies I do exploration and tracking exercises with are all domesticated. The majority are from the Netherlands and some are from other parts of Europe. I work with these horses in temperatures ranging between -10° – 30° Celsius. These are horses and ponies of all sizes and breeds. Some do not have pedigree papers. Their ages range from 3 to 30 years old.

A small number of the horses are housed in a paddock paradise, but most live in more traditional stables in which the horse is at pasture during the day and spends the evening and night in an individual stall. The horses do not work in a riding school; each has one or two owners. The majority of the horses are ridden recreationally, in dressage, hunt seat, or Western style. A smaller number ride in competitions, sometimes at a higher level.

When I talk about horses and ponies, I am referring to domesticated horses. I also use the word *horse* for both horses and ponies. I refer to all horses as *he*, unless it is clear the horse in question is a mare. I refer to all humans as she, unless it is clear the human is a male.

When I speak of the horse generally in a singular form, it can also be read as 'horses' in the plural. Also, I try to speak plainly, using everyday language, as much as possible. For instance, I will sooner say 'a reward that motivates and stimulates' than use the term 'reinforcement'.

When I talk about teaching a horse to track, it is a reflection of the fact that many domesticated horses no longer instinctively use their noses to track. However, I find it condescending to speak of 'teaching' a horse to track. After all, this is a skill every horse possesses. I am letting them rediscover the ability, so that they can start using it again. However, for the sake of convenience, I use the word 'teach'.

REFERENCES

I like a connection between theory and practice. I like to investigate whether the experiences I have in the many exploration and tracking sessions I do can be biologically explained, which allows me to understand why something does or does not work from a biological perspective. The same applied when I wrote this book. I started writing with one figurative foot in the sand and one in books. I chose to give basic information about the brain and some neurotransmitters and stress hormones. Of course a body consists of more than just these neurotransmitters and stress hormones, so a caveat about my work is that I do not do justice to all the other components that also influence the body and that are so much more complex than my representation. I hope, however, that my explanation offers you a basic framework from within which you can understand your horse's behaviour and the developments he is going through from a biological perspective. I do know, though, that much more could have been included and remains to be explored.

I was not always able to literally find my experiences in the books and studies I had. Where this is the case, I note this clearly and label my thoughts as assumptions and reflections. (If there are studies that fall within the scope of this book that I have overlooked that you would like to direct my attention to, I would love to hear it, of course. Then I can integrate them into any possible future editions of this book.)

The books and studies that have been of significant help to me are listed in the references and bibliography. This implies that I did not make note of everything I read. In deciding how to use the information I gathered from these and other sources, I looked for the correct balance. The question of how much information I ought to share in certain places was one I regularly asked myself. There is so much to say! I hope I have found a proper balance, in which I do not go into too much depth theory-wise but still provide enough background material for you to be able to place the things I say about the practice in a theoretical framework. If you want to know more, the books in the bibliography might inspire you. In this book, I have tried to name the sources from a single book, so that if you want to buy them, you do not have to purchase so many different sources. I have also tried to find sources that are easy to find or even free on the internet in full-text PDF form.

Wherever possible, I use information that emerged directly from research conducted with horses by other scientists or myself. However, I also include studies that focus on the olfactory systems and abilities of humans, dogs, rodents, rats, mice, and fish because a lot more research and knowledge is available about these subjects. Every species experiences the world and the stimuli in it in its own way. A scent that might attract one species can drive away another. However, when we look at how different mammal species processes scents and stimuli at the level of the brain, there are mostly similarities. Neurologist Jaak Panksepp aptly said, 'Most vertebrates share the fundamental structure of the olfactory system, as well as its mechanisms, and the homogeneity within the mammalian brain is truly impressive' (Panksepp 1998; Panksepp and Biven 2012).[1] I can only imagine that more research will be done into horse's sense of smell. I am already looking forward to contributing to it.

REFERENCE

1. Gadbois S., Reeve C. (2014) *Canine olfaction: Scent, sign, and situation.* https://www.researchgate.net/publication/280446218_Canine_Olfaction_Scent_Sign_and_Situation,5.

Part 1: Exploration

1 *Introducing exploration*

SCENTS, sounds, tastes: Living in an environment designed by people for people is different for every horse. It is like a game of chance, in which everything depends on the turn of a card or a roll of the dice. He cannot choose where or how he will live. That depends on many factors, such as his heritage, suitability to various sports, height, physical appearance, character, or the wallet size of the person who buys him. Wherever the horse ends up, his day-to-day life depends greatly on his owner – that person's choices with regard to where the horse is housed, how much time he gets to spend at pasture, whether there are other horses at pasture with him and which ones, as well as his owner's ambitions and goals, choice of work intensity (how often, how much), and whether he has to travel or not. The point is clear: each individual horse's life is like a puzzle that can be laid in uncountable different ways. The same is true of the stimuli with which he is confronted every day. There can be many or few, always the same ones or always different ones. There can be days with many stimuli, interspersed with days that contain few. These can be cars or mopeds driving by, leaving a gasoline scent; tractors coming up from behind while riding in the country; the mailman delivering a package; the scent of cows released in a neighbouring field; the sound of a child crying; an air conditioner that is always buzzing in the background; an escaped dog running around the pasture; planes flying overhead; back pain due to an awkward step at pasture that is unnoticed by his rider or handler; and being touched by one person or by many. To the horse, living in a human world means living in a world filled with stimuli people maintain and ignore, including some that cannot be or are not changed.

The degree to which the horse can handle the stimuli in his living environment and how he does this determines his wellbeing to a large degree. It influences his physical and mental health, his emotions and behaviour, the degree to which he can fulfil his task, and how well he learns and remembers. It has an effect on the relationships he builds and maintains with people. If a horse cannot or largely cannot handle the stimuli in his environment, the chances are good that he will end up in a negative spiral. And if something goes wrong, it is usually the horse who pays the price.

That is why, in this book, I am focused primarily on the mental development of the horse. What does it take to train him in such a way that he can handle stimuli from our human world so that he feels comfortable around them, forms positive associations and feelings, and feels good mentally and physically?

1.1 THE COMFORT MAP

When you want to create a programme for your horse and his mental development, you cannot escape having to establish a baseline, seeing which stimuli your horse encounters in his life. That way, you also have a good calibration point that you can use for comparisons in the future.

STEP 1: CHOOSING A PERIOD OF TIME OR CATEGORY YOU ARE GOING TO CHART

This can be a day, an evening, a time when there are no other horses and your horse is alone, the day before and after a competition, at pasture, in the grooming area, in his stall, etc. The more time periods and categories you pick, the better you will be able to chart the stimuli your horse deals within his life. Stimuli can include sounds he hears; scents he smells; or stimuli he sees, tastes, or feels, either from his environment or from inside himself. It is clear that we cannot identify all these stimuli; unfortunately, our senses are not as acute as a horse's. It is also hard for us to know what stimuli your horse is feeling on the inside. These can be things such as hunger, thirst, or pain before a mare goes into heat, which can translate into irritability when being touched.

STEP 2: LISTING YOUR HORSE'S BODY FEATURES AND BEHAVIOURS WITH REGARD TO THESE STIMULI

You can categorize these body features and behaviours into different colour zones; I use green, yellow, orange, and red. Green is the colour of relaxation; the horse has no problem handling the stimulus/stimuli/situation. Red, on the other hand, symbolises the other end of the spectrum in which the horse wants to flee from the stimulus/stimuli/situation or chase it away. Yellow and orange are the stages in between.

In the Appendix, I give descriptions of horses' body features and signals in the different colour zones. There are also two communication ladders and an empty note-taking page you can use. That way, you have some tools to help you establish your horse's zones.

Here is a comfort map for my horse, Vosje. On this map, I have written some stimuli he encounters when he is at pasture (**Fig. 1.1**).

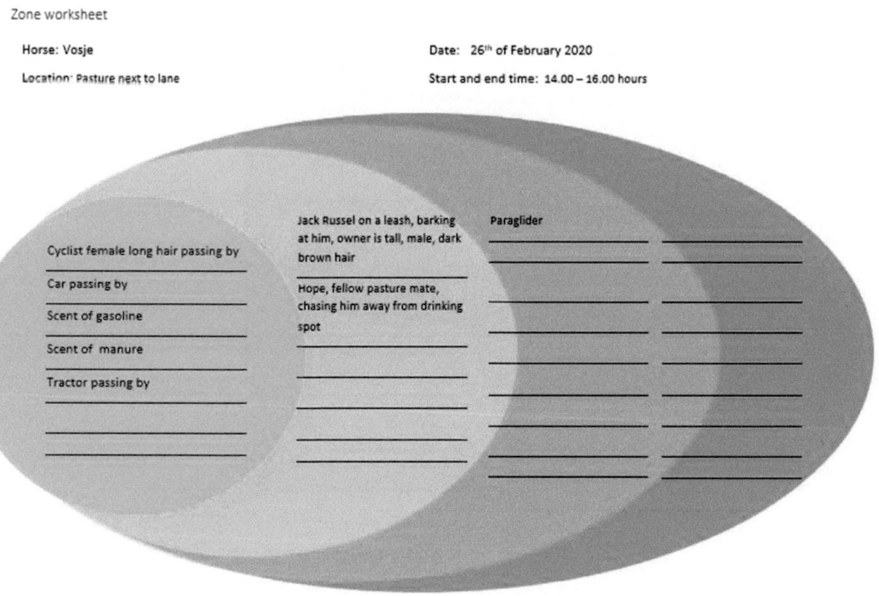

Zone worksheet

Horse: Vosje

Date: 26th of February 2020

Location: Pasture next to lane

Start and end time: 14.00 – 16.00 hours

Cyclist female long hair passing by

Car passing by

Scent of gasoline

Scent of manure

Tractor passing by

Jack Russel on a leash, barking at him, owner is tall, male, dark brown hair

Hope, fellow pasture mate, chasing him away from drinking spot

Paraglider

Fig. 1.1

SEPARATION ANXIETY HAS ITS OWN COMFORT MAP

When I am dealing with a horse who has separation anxiety, I create an alternative comfort map. I print out an areal picture of the location (Google Maps) where the horse is housed, and I colour-code the areas according to what the horse can handle when, for instance, he is taken away from the other horses (**Figs. 1.2, 1.3**).

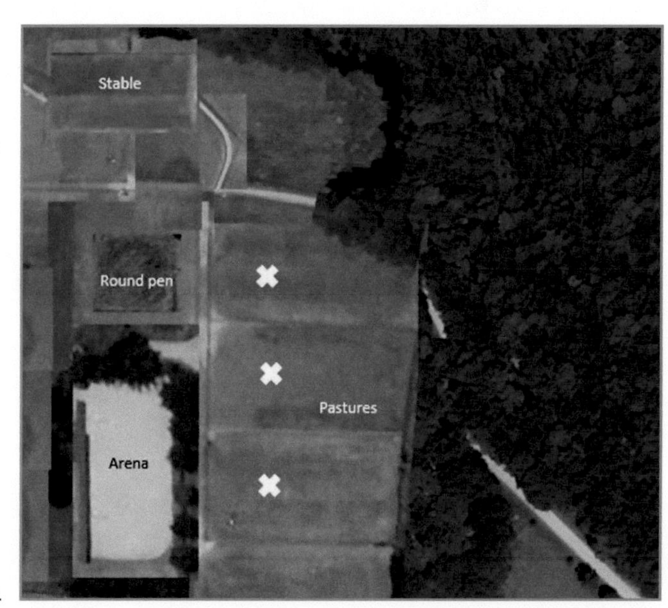

Fig. 1.2 Patricia, Esther's horse, is at pasture (the area with the three crosses) with other horses. When Esther comes to collect her from pasture, Patricia has trouble with this. We chart how far Patricia can be removed from the other horses. The primary stimulus we are focused on now is the distance between Patricia and the other horses. Other stimuli can be added to this, and you can chart them, too.

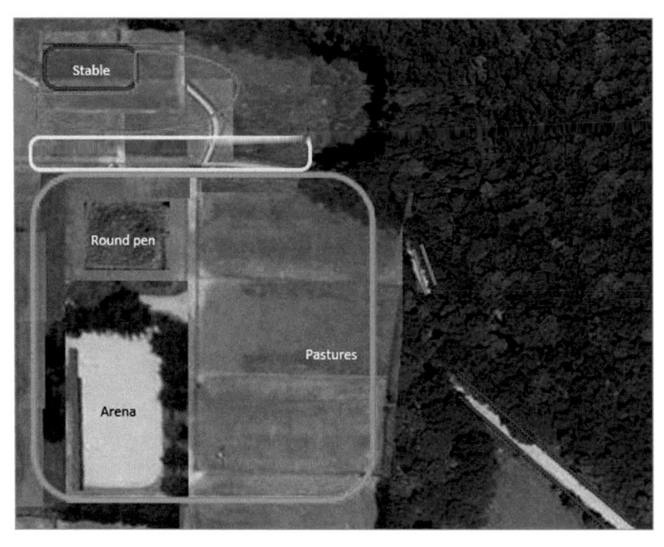

Fig. 1.3 By looking at Patricia's body features and behaviours, we charted these zones. We will use this information during our exercises. She also serves as a good point of comparison. Patricia's case is described in more detail in Chapter 4.

It can also be interesting to create comfort maps for yourself and see which stimuli go in your own green, yellow, orange, and red circle to see which zones overlap with your horse's and which do not. That way, you can use the information in your shared lives and work. If you are riding in the country with your horse and he is in his green zone, but you are in your yellow or orange zone, you can give the horse a bit more space and lean on him a little and vice versa. There will be more about this in section 9.4.

SHUTTING THINGS OUT IS NOT THE SAME AS BEING COMFORTABLE

Pay extra close attention to the horse who obeys, acquiesces to the handler or rider's wishes, and calmly undergoes stimuli or situations. Of course, it could be that he is really able to handle these stimuli. That would be great. Sometimes, however, it happens that a horse shuts himself down for shorter or longer periods, without us humans noticing it. The moments or longer periods during which he shuts down are characterised by stillness and passivity of motion, a horizontal to mid-low or low head-neck position, and an inward gaze, possibly with half-closed eyes. There is little facial expression, in this case, and a stillness in the face (you will not see any wrinkles either). These moments of shutting down can occur within a calm movement pattern, but also within a more energetic one.

People often tell me they find it hard to recognize this shutting down. My answer: It is hard! A horse can present the same picture when he is dozing or in pain. A good tool to use here might be the questions: 'Is this horse watching TV or not? And should that not make more sense right now?'

I like the watching TV analogy because it lets you feel and experience for yourself which elements you are looking for in your observations. When you are watching TV yourself, you are relaxed. Your gaze is turned outward, but your head and neck stay in their relaxed posture. This is different from shutting down, in which case you turn inward, as well as from a state of tension, in which your gaze is turned even more outward, but now it is no longer relaxed but tense. In that case, your eyes open more widely, there is a movement of weight and energy forward, and your head and neck tense up a little and tend forward a bit.

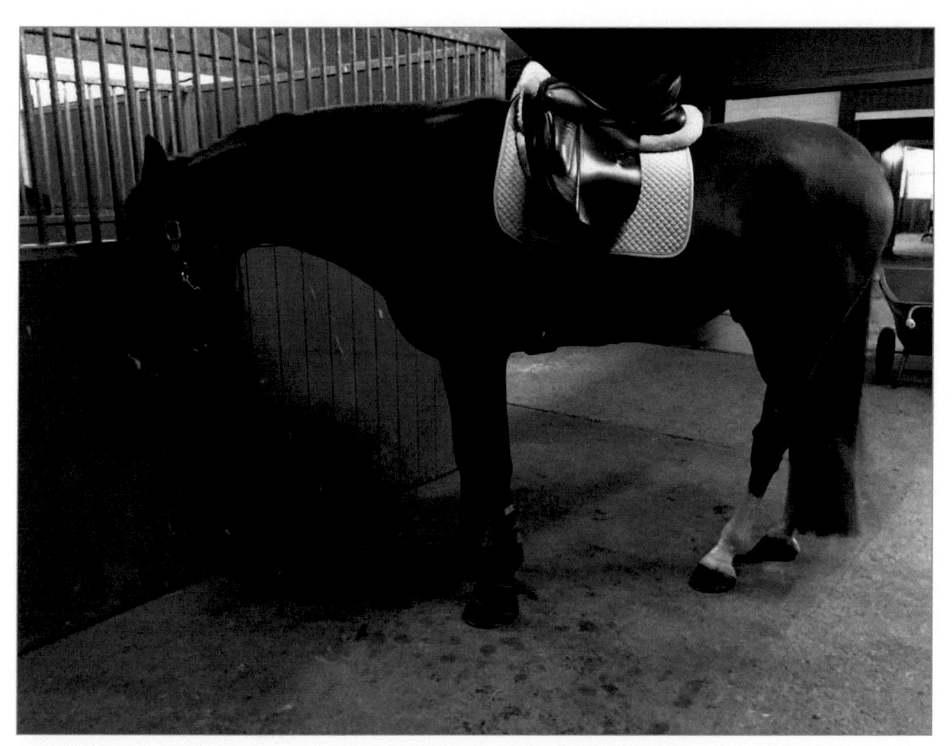

Fig. 1.4 This 4-year-old horse is housed at a large training stable. People are constantly moving back and forth, some leading other horses or pushing wheelbarrows. This horse's posture indicates shutting down. You would hope and expect a 4-year-old horse to be peering around as if he were watching TV. This horse stood like this for half an hour until a rider came to finish saddling him. In case of a longer shutting down, like here, you can also be dealing with fatigue, pain, or depression. It would also be a good reason to take a good look at this horse's life and see what could be improved.

A horse who shuts down is easier to handle for people. It is important to realise, however, that this horse also runs health risks associated with stress-related problems (see section 2.2).

If you have charted the stimuli and situations that apply to your horse, his life, and your shared lives, then you can keep these in mind when you start doing exploration and tracking exercises. That way, wherever possible, you can adapt these to the specific stimuli your horse has to get used to. Is your horse always in the green zone? Do not let that stop you. This horse will also get a lot of use and fun out of exploration and tracking.

1.2 EXPLORATION AND EMPOWERMENT

I believe that we, horse owners, have an obligation to our horses to help them to understand the world around them and to cope with the stimuli they encounter. This is in addition to the obvious obligation we have to meet the horse's needs, which is reflected in 10 freedoms, the first five of which include: freedom from hunger and thirst; freedom from discomfort; freedom from pain, injury, and disease; freedom from distress and fear; and freedom to express natural behaviour. These five are often seen as the basics. However, the next five, which Marc Bekoff originally established for dogs, are just as important for and applicable to horses: 'the freedom from avoidable or treatable illness and disability, the freedom to be themselves, the freedom to exercise control and choice, the freedom to frolic and have fun and the freedom to have privacy and safe zones'.[1] This also includes a life among and with other horses.

The greater your horse's comfort zone in his work and life, the better he is able to handle his life. There is no upper limit to this. There is so much to experience and discover. There are so many benefits to be had for a horse with a very small comfort zone, a 'bomb-proof' horse with the huge comfort zone, and all the ones in between.

My exploration and empowerment method rests on two pillars:

1. Make the horse independent and empower him. The emphasis has to be on the horse as an individual not as a part of the horse–rider or horse–handler combination.
2. Stimulate the horse's impulse to explore. His seeking system is activated and/or further developed.

PILLAR 1 – INDEPENDENCE AND EMPOWERMENT

Everybody knows that special older horse who has seen so much and whose responses to everything are so balanced that a young child can take him for a walk. It is not that the child is making the horse comply. No, the horse can handle life on his own. He knows the stimuli and can stand on his own four feet, and being treated as such is what many horses need. Is it not strange, after all, that we treat every horse the same, regardless of their age? Whether he is 3 years old or 20, with a wealth of experience behind him, we stick to our habits, including how we tie him up, how we lead him on a walk, and how long we groom him.

What we find easy to do with children, we find much harder to do with horses. If we have children, we prepare them for a life of independence. As they get older, the things they have to do change as does our involvement in their lives. We let children go through processes step by step, from putting a few groceries on the conveyor belt one day, to putting them into the bag another, to paying the cashier, to getting bread while we wait outside the store, to finally doing the groceries on their own.

This goal of making the horse more independent in the human world is what I strive for. Of course we have to work within the margins of ability and safety. If you have a 4- or 5-year-old child in a big playground, you do not just let him loose either. First, you check out which playground equipment he can handle, give some tips, help out here and there, and you give him freedom to experience in areas where it is possible. When it comes to older kids, say 10 to 12, you do not need to do this check beforehand. That is how you calibrate. You do the same thing with your horse. You are constantly creating a custom package. I do see in others and in myself that letting go can often be a bit more difficult with horses.

I was walking down a forest path with Vosje. He was in front, deciding where we went. We came to the edge of the forest. There was a barrier across the broad path we were on, blocking our way. To the left of the barrier, there was a very narrow path that curled around it. There was a big puddle of water in the middle of this narrow path, and electric wire running along its other side, marking the edge of a cow pasture. My initial, instinctive reaction was to stop Vosje, investigate the path myself first, and then lead him through it. I stopped myself though, thinking, 'Vos is over 25 and familiar with electric fences, water, and barriers: He should easily be able to handle this himself'. And, indeed, he passed through it without difficulties.

It is human behaviour to allow the horse little or no freedom to act on his own. This is often a product of our love for the horse and wanting to take care of him. It could also be the learned notion that we humans always have to be in charge. However, these habits keep the horse in a dependent role and far less able to fully develop as an individual, something that brings enormous benefits for the horse and us.

It is important to note that this is a fundamentally different premise than that of other programmes that train the horse to handle stimuli, mostly by teaching them to endure them. These are two common examples:

A horse is in the arena. He is wearing a halter or bridle. The trainer holds him as he or a helper confront the horse with stimuli. This can be a flag being drawn across the horse's back, or a saddle blanket (if he is not used to this), or objects that are held that make noise. In response, the horse will show flight behaviour. This can be of various gradations: from a slight movement of his weight away from the stimulus or taking a few steps away from it, to a stronger reaction, in which the trainer has to hold the horse firmly because the horse really wants to escape the situation. The moment the horse starts to show behaviour the trainer sees as positive in response to the ongoing stimulus, the trainer stops applying the stimulus.

A 'startle training circuit'. The arena is overflowing with objects. The horses wear halters or bridles. Their handlers lead them into the arena on foot or ride them. The horses are lead or ridden in between the objects. Sometimes an umbrella is opened and closed a few times. People rattle things. The horses are not allowed to smell the objects or anything. After a certain time, regardless of the emotions or behaviour the horses are showing, everyone leaves the arena again: the trainer's hour is up.

Disadvantages of these methods:

- The horse undergoes the stimuli. This does not accurately demonstrate his own motivation and ability to approach the stimulus, to want to discover it and interact with it. (A horse who is curious about a stimulus and wants to experience it shows different body features and gives different signals than a horse who is undergoing it.)
- It focuses on the behaviour instead of the underlying emotion. The degree to which a horse obediently stands still because that is what he was taught to do is not necessarily a reflection of an equally calm underlying emotional and physical state. In that case, the horse could be standing still as a coping mechanism, while still experiencing tension at the same time.[2] This could increase the chance that the horse will continue to show startle responses in the future or even show heightened fear responses to the practiced stimuli or to the stimuli he has encountered during this learning experience (such as scents, sounds, movements the people made, etc.) because, to him, the underlying emotion was negative. If the horse feels a negative underlying emotion, it can have a detrimental effect on his mental and physical health.
- If a specific rider or handler taught the horse to display a certain desirable behaviour while the underlying emotion is still negative and the tension is high, there is a chance the horse will only show this desirable behaviour in response to the stimulus in question in the presence of the person who taught him this. In the company of people other than that specific handler or rider, tension features and behaviours can still be expressed.
- If the handling and timing of the rider or handler is bad, there is a chance that the horse builds up even more tension with regard to the stimulus, and the relationship between the horse and the person deteriorates.

PILLAR 2—THE SEEKING SYSTEM IS ACTIVATED OR STIMULATED FURTHER.

A curious brain is a healthy brain.

Turid Rugaas

In this method, curiosity is aroused, and the horse is stimulated to investigate, to experience, and to do for himself. This sets in motion mental and physical processes that have a positive effect on the horse's wellbeing, emotions, and behaviour.

1.3 THE BENEFITS OF EXPLORATION AND EMPOWERMENT

Exploration and empowerment can be done in facilities of different kinds and sizes. You can also tailor the activities to your horse's needs and abilities. Regularly applying the empowerment and exploration method based on the needs and abilities of your horse has the following advantages:

- It meets your horse's natural need to search and explore.
- It meets your horse's need for independence and freedom of choice.
- Because the horse investigates stimuli on his own, he is not influenced by others, which enables him to build an individual frame of reference and go through an individual process of development.
- Because the horse investigates the stimuli and is given the freedom to do this in his own way, for instance by tasting, smelling, hearing, seeing, or feeling, he experiences the stimuli. He gets to know and understand them. He can place them within his own frame of reference.
- Because the horse investigates stimuli at his own pace and according to his own ability, there is a good chance that he will create positive associations with the stimuli.
- Discovering and investigating stimuli brings increased joy in life and an eagerness to undertake activities.
- It brings an increased eagerness to try new things and explore further.
- Because the horse is in control of his own pace of exploration, there is a greater chance that he will remain within a tension zone he can handle or experiences only light tension.
- It causes the horse to practice self-regulation, enabling him to calm himself when he is experiencing light tension. Because he controls the pace of exploration, he is the one who gets to overcome slight hesitations, and he practices doing this.
- Because the horse is in control of the pace of exploration, increasing the chance that he will remain within a tension zone he can handle, and because is allowed

to explore in his own way, there is a good chance the information will be stored in his long-term memory.

- The fact that the horse is in control of the pace of exploration gives us, the rider or handler, valuable information about the horse.

If you do it regularly (for instance, twice a week on average), exploration and empowerment translates into:

- A decrease in fear and aggression responses with regard to known stimuli.
- A decrease in fear and aggression responses with regard to new stimuli.
- A decrease in impulsivity.
- A decrease in overreactions and tension.
- A decrease in frustration (which had been caused by not being able to execute natural seeking behaviour).
- A decrease in boredom.
- A decrease in the development of chronically elevated stress levels.
- A decrease in the chance of developing 'shutdowns', learned helplessness, depression, and lethargy.
- An increase in impulse control.
- An increase in the amount of time in which a horse can concentrate.
- An increase in problem-solving ability, or an increase in displaying problem-solving ability.
- An increase in generalisation. If he understands a logical concept, he can link comparable concepts to it. A horse who generalises and is used to a long-haired black dog will more easily become accustomed to a short-haired grey dog, because he is familiar with the dog concept. A horse who does not generalise has to get used to the short-haired grey dog all over again.
- An increase in long-term memory capacity.
- An increase in happiness and joy in life.
- An increase of calm in the horse's behaviour and nature.
- A healthy immune system and a lower chance of developing stress-related ailments.
- Better body control.
- An increase in seeking out and maintaining social relationships.
- An increase in engagement with the rider and/or handler. It improves the relationship between horse and human from the horse's perspective.
- An increase in comfort with regard to the presence of other people.
- An increase in willingness when it comes to the tasks people ask him to perform because of the development of reciprocity.

REFERENCES

1. Bekoff M. (2019) *Dogs, captivity, and freedom: Unleash them whenever you can.* https://www.psychologytoday.com/us/blog/animal-emotions/201903/dogs-captivity-and-freedom-unleash-them-whenever-you-can
2. Squibb K., et al. (2018) Poker face: Discrepancies in behaviour and affective states in horses during stressful handling procedures. *Applied Animal Behaviour Science* 202:34–38 https://www.sciencedirect.com/science/article/abs/pii/S0168159118300674

2 *Brain and biology: What explains these changes?*

IN ORDER to explain the effect exploration and tracking have on the horse, we have to dive into the world of biology. If you already know the biological explanation or if biology really is not your thing, skip ahead to section 2.3. If, however, you would like some more information about this, read the following section about the brain, some neurotransmitters, and hormones. That will give you a biological framework in which to place the benefits brought by exploration and empowerment and tracking.

2.1 ABOUT THE BRAIN AND NEUROTRANSMITTERS

Nothing is more mind boggling than thinking about the brain, the control centre of our bodies. It contains our individuality and determines how we feel and behave, how we move, when we become tired and hungry, how motivated and resilient we are, how we remember and flee, to whom we feel sexually attracted, and so much more. Although we think in pictures and language, that is not what rules our heads. No, it is uncountable forms of electricity. And where these electric charges form, how they spread, how they are reduced or amplified on the journey, whether they go left or right are the factors that determine the functioning of our brains. If our brains change, our emotions and behaviours change. And the reverse is also true: a changed pattern of behaviour stimulates changes in the brain as well.

According to estimates, the mammalian brain consists of 100 million brain cells/neurons. Every brain cell consists of dendrites (with an average of 10,000 dendritic spines per neuron[1]), a cell body, an axon, and axon terminals (also about 10,000[1]). Every neuron carries its own electrical charge, which is not constant but variable. Around every brain cell, there are glia cells (not shown in the illustration). Glia cells fulfil different functions, among which are making sure the neurons get nutrients and producing myelin, the fatty substance that surrounds an axon and that plays a role in the conductivity of electricity. They also clean up cell debris when there is a neural injury or cell death, maintain the firmness of the brain tissue, and keep groups of neurons apart from each other. In addition, some glia cells tell axons in which direction to grow (**Fig. 2.1**).[2]

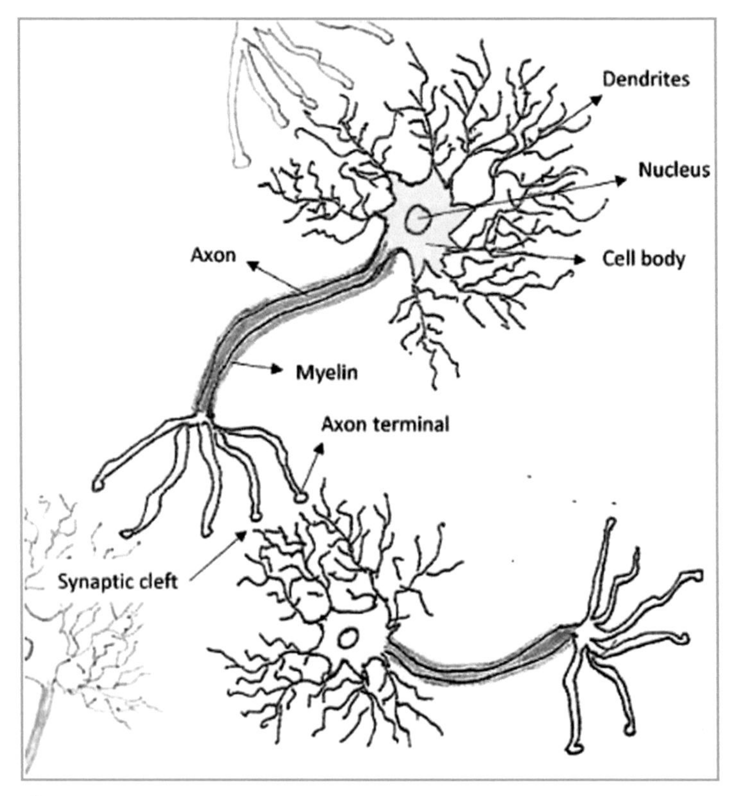

Fig. 2.1

The illustration here is far from life-size. To give you an idea of the size and proportions here, the diameter of a human hair is 50 to 100 micrometres. An axon is 5 to 10 micrometres. A dendrite is just a few micrometres in diameter. A cell body can be 5 micrometre to 1 millimetre.[3] *The synaptic cleft is 0.03 to 0.05 micrometres.*[3] *(1 millimetre = 1000 micrometres).*

The brain cells communicate with other brain cells by means of electrical pulses for which the dendrites function as receivers and the axon terminals as transmitters. The degree to which the sent electricity influences the electrical charge the neuron already has and, literally, the path it then takes, causes a reaction. As you can see in the illustration, however, brain cells are not directly connected to one another, so if the electrical charge did not receive help in making the crossing to another brain cell, it would not be able to do so. This help comes in the form of neurotransmitters. They are the messengers who make sure electricity makes it across the synaptic cleft and to the other cell.

Neurotransmitters are produced inside the brain cell. When they are done, they are kept in a small sack close to the synaptic cleft until they are needed.

The moment a neuron receives an electrical charge from another cell through its dendrites, which it will then send further on its way, the electrical signal travels from the axon to the axon terminal. The moment the electrical signal reaches the axon terminal, a neurotransmitter is released. This neurotransmitter blends in with the electrical signal, and together they cross the synaptic cleft. The neurotransmitter then attaches itself to a specially suited dendrite of the next cell, allowing the electrical charge to continue on its way. However, the electrical charge will no longer be the same. The neurotransmitter attaching itself to the dendrite has enhanced or reduced the electrical charge, so it is different when it continues on its way through the dendrite of the next brain cell. When the neurotransmitter is done, and what remains of it is either left inside the synaptic cleft where it is decomposed, or it is re-imbibed by the neuron it came from and used again.

The neurotransmitters I regularly refer to in this book are adrenaline, noradrenaline, dopamine, and the peptide neurotensin. Adrenaline and noradrenaline are neurotransmitters that provide a strong boost to the electrical signal they guide across the synaptic cleft. However, their chemical effect does not only take place in between brain cells. Adrenaline and noradrenaline influence and stimulate every cell in our bodies that is susceptible to them.

Dopamine is a neurotransmitter that is directly involved in the feeling of reward and wellbeing. However, it is not the case that these neurotransmitters mostly become active when a reward is being given. In fact, the expectation of a reward also releases dopamine. That is why dopamine plays a strong motivational role. It causes goal-directed behaviour. A study on dogs showed that breeds that have a lot of dopamine receptors, which allows them to process large quantities of dopamine, are the same breeds that are very willing to work and highly persistent, such as the border collie, the Belgian Malinois, and the Jack Russell terrier.[4] Dopamine almost has an addictive quality. A dopamine shortage is linked to depression and an inability to feel joy.[1] Anxiety and acute and chronic stress make it more difficult, if not impossible, for dopamine receptors to do their work.[1]

Neurotensin is a neuropeptide that turns on the seeking system. In their book *The Archaeology of Mind: Neuroevolutionary Origins of Human Emotions*, Panksepp and Biven wrote that animals are normally enthusiastic about acquiring neuropeptides that turn on the seeking system and dislike chemicals, such as dynorphin, that deactivate it. The moment dopamine shortages start to occur, and the aversive feelings produced by dynorphin begin to prevail, people feel depressed.[5]

2.2 STRESS HORMONES

Stress hormones that I will discuss more often in this book are adrenaline, noradrenaline, and cortisol. And yes, you are right, we just talked about adrenaline and noradrenaline as neurotransmitters; however, adrenaline and noradrenaline are multifunctional, as they can also be excreted by the adrenal glands as hormones. This can happen when a horse considers a situation or stimulus to be exciting or threatening. This can be a situation or stimulus the horse is experiencing at that moment, or what is happening in the moment can be reminding him of a dangerous or exciting situation from his past. If this is the case, the amygdala (the clusters in the brain that play a major part in the regulation of fear) stimulates the hypothalamus, which in turn stimulates the pituitary gland, which in turn stimulates the adrenal glands to secrete adrenaline and noradrenaline. Adrenaline and noradrenaline enable the body to flee or fight at full power that instant. There is an increase in blood pressure, breath, and heart rate, and more blood flows to the muscles. Adrenaline and noradrenaline are drivers of this. Because the body is unable to maintain this state for long, after a few seconds or minutes, the adrenal glands release the hormone cortisol. Cortisol enables the body to keep up a strenuous effort for a longer time. Once the exertion passes, the body can relax again. If the period of exertion lasts longer, however, or if the horse is in a situation in which he is exposed to a lot of moments of stress from which he cannot recover, the cortisol will continue to play a decisive role in his body. Cortisol enables the body to keep up the energy required. This is very useful, of course, but there is a downside: It does this by pilfering energy from other processes that use energy.[6] This puts pressure on the immune system, which deteriorates the longer the cortisol is released, which may lead to reduced health.[7, 8] But there are other disadvantages, which I have also listed in my previous book:

- Reduced functioning of the hippocampus: cell activity decreases, there are fewer synapses, and there is less information transferral, which leads to reduced cognitive learning capacity, a reduced ability to solve problems, and reduced memory capacity.[7, 5, 9]
- The threshold at which the horse responds fearfully lowers. The horse is startled more quickly and more often, now also by things that never used to startle him.[7, 5]
- Increased risk of stereotyped behaviours.[10]
- Increased risk of developing injurious behaviours such as self-mutilation and increased aggression.[8]
- Increased risk of developing lethargy and depression.[11, 5] This happens because, in time, the adrenal glands become exhausted.
- Poor quality of sleep.
- Decreased fertility.[8]
- Gastric disorders and damage.[8]
- Possibly a more difficult cooperation with his handlers and riders.
- Lacking the feeling of homeostasis: no balanced sense of wellbeing.[12]

NERVES

And, finally, how does all this information get from the brain to the muscles and organs? And how do the muscles and organs communicate with the brain, so that constant coordination is possible? This job is done by the nerves.

2.3 EXPLORATION AND EMPOWERMENT FROM A BIOLOGICAL PERSPECTIVE

As a behaviour consultant, I have been helping people, dogs, and horses since 2003. Stimulating the thirst for exploration and tracking has become a staple in my lessons and treatment plans (you can read more about this in section 4). In practice, I could see the many benefits named in section 1.3 happening before my eyes. As I wrote in the introduction, I am also interested in the biological background of these benefits. What processes in the body are responsible for this? Here, I will discuss a small number of them.

Letting the horse explore, have new experiences, and solve problems (I consider enriched environments and tracking ways to do this) causes changes to take place in his brain that are very beneficial. Some of these include:

- Stimulated brain cells create more connections with other brain cells. The dendrites acquire more bifurcations, and the axon terminals grow new sprouts that might grow in a different direction.[1] This is beneficial because the more neural connections you have, the more you can use them to place, understand, and remember all sorts of stimuli you encounter. New and unfamiliar stimuli will also be understood much more quickly and can be placed within the neural network and frame of reference that is already present.[1]
- Brain cells that are often used are myelinated or further myelinated. This means that myelin, a fatty substance, is wrapped around the axon or additionally deposited. This makes the axon better insulated, reducing leakage of the current out of the axon. This enables more information to be transmitted to other cells and at an even higher rate.[13]
- New cells are produced in the hippocampus over the entire lifetime.[1] The hippocampus is viewed as the place where memory is stored. It plays an important role in our thoughts and actions. The stored memories enable a connection to be made between experiences and thoughts from the past and those happening now, allowing you to include experience and knowledge from the past in your current life.
- The urge to track stimulates extra dopamine secretion in the body.

HORSES WHO NEED MORE CARE

The benefits above apply to all horses, but for horses who need extra care, they are especially valuable. That is why I want to take a moment to discuss these horses from a biological perspective. These can be horses who have lived through or are still experiencing many moments of acute stress or are chronically stressed. There can be many reasons for this. Maybe they have trouble being alone and have separation anxiety. The horse could be scared of certain stimuli (including people), making him hard to handle. It could be a horse who bites, cannot be collected from his pasture, does not want to go into the trailer, startles a lot, and flees the scene. It could also be a horse who is in pain. These can also be horses who shut themselves off from the stimuli around them for a shorter or longer time, causing depression if it does not let up.

On the basis of his body features and behaviours, or the lack thereof, I can deduce a horse's level of tension or the degree of shutting down. For more information about body features and signals of horses, have a look at my book *Language Signs and Calming Signals of Horses.*

Although the very tense horse and the horse who is shutting himself down present a very different picture in terms of behaviours and body features, they are very much alike when it comes to exploration and tracking. When I offer them stimuli in a suitable enriched environment (more about that in Chapter 3), they tend to deal with this differently than horses who are mentally and physically healthy. Both the horse who has shut down and the very tense horse need longer to start investigating objects (of course, I try to use objects that align with their abilities as much as possible, so the horses can have experiences of success).

The horse who has shut down can withdraw into himself, with an inward gaze and a head and neck position between mid-high and mid-low. He becomes still, shows few movements, takes little to no initiative, seems more withdrawn, and he is definitely not watching TV (see section 1.1) when it comes to the objects. He does not approach the objects, or, for that matter, do anything with any level of energy, such as seeking contact with people, whinnying at other horses, and eating grass. This horse lacks the energy to do anything that is outside of his normal routine. If you put this horse inside an enriched environment, he will just stand there, looking tired and passive. This can last for longer (sometimes several sessions divided over 2–3 weeks) than a horse who is tired or does not know he is allowed to explore. The latter horse might develop alternative strategies such as for instance heading for the exit or starting to eat grass if it is there.

A horse who has a lot of moments of acute stress or who is chronically tense presents an entirely different picture. He is very energetic. He is not calm enough to investigate the objects or track. His body is prepped to want to leave the situation. For example, he might start to pace in front of the exit if he cannot leave, he may defecate more often, or eat in a very rushed way. If his tension level rises further, he will no longer eat or drink when you offer food or water. You can also see other body features, such as tensed muscles, rounder eyes and nostrils, a tense chin and lips, and a distortion of the nose and mouth.

When I make a behaviour adjustment plan for both types of horses, adding tailor-made exploration and tracking exercises, I get the impression that the horses who do more of these exercises reach healthy homeostasis more quickly than horses whose handlers offered this less or not at all. The horse who is shut down is increasingly interested in his environment and starts to take initiative more quickly. The tense horse becomes calmer and more balanced, with fewer startle responses. The benefits discussed in section 1.3 are also within their reach sooner. When I connect this to the biology underpinning it all, this information stands out:

- Frequent moments of acute stress or chronic stress cause an elevated glucocorticoid level (this includes cortisol) and a lower dopamine level.[1]
- A dopamine shortage is linked to depression and the inability to feel joy.[1]
- Frequent moments of acute stress or chronic stress cause a reduced functioning of the hippocampus: Cell activity decreases, there are fewer synapses, and there is less information transferral, which leads to reduced cognitive learning capacity, ability to solve problems, and memory capacity.[7, 5, 9]
- Frequent moments of acute stress or chronic stress create more excitable synapses in the amygdala.[1] This point, in combination with the point above, explains why a horse who lives with many moments of acute stress or chronic stress suffers from stress-induced impulsivity and poor emotional regulation.[1]
- Neurotensin is a neuropeptide that turns on the seeking system. Animals are normally enthusiastic about acquiring neuropeptides that turn on the seeking system and dislike chemicals that deactivate it.[5]

EXPLORATION AND SCENTWORK AS A TOOL AND A MEDICINE

What exploration and tracking do very explicitly is activate the thirst for exploration. This involves the entire body and all the senses. When we are talking specifically about tracking, the nose is additionally stimulated. Exploration and tracking stimulate the olfactory tubercle, which is densely wired with dopamine receptors.[14] This is also largely true of the pyriform cortex.[4] These areas of the brain are activated during exploration and tracking, stimulating the secretion of dopamine in the body of the horse, more than would normally happen without exploration and tracking exercises.

In a horse who is dealing with shutdowns or depression, this extra stimulation can lead to the elevation of their inadequate dopamine level. After all, you are giving it an upward boost toward normal levels, so much so that the horse is able to reach healthy homeostasis more quickly.

In a horse who lives with many moments of acute stress or chronic stress, the stimulation of dopamine release causes his low dopamine levels to get an upward boost, enabling him to achieve healthy homeostasis more quickly as well. In addition, the stimulation of activated dopaminergic reward pathways causes an inhibition of the release of the corticotrophin-releasing hormone, which also decreases the secretion of cortisol, giving it a double positive effect.[1]

It is also easy to imagine that, for a horse who tracks or does exploration sessions more often, it will get easier and easier to turn on his 'urge to explore'. With these exercises, we stimulate neurotensin, which plays a role in the activation of the seeking system and can cause a positive feeling in the horse.

For me, exploration and scentwork are *the* tools and mental medicine that stimulate the horse to achieve and maintain healthy homeostasis: the state in which the body is physically and mentally in optimal balance, even if changes occur in the environment. That exploration and tracking help the horse to achieve healthy homeostasis does not surprise me. Exploring stimuli and tracking for food, for instance, is so essential to the horse's survival that it is only logical that the horse's body would activate all sorts of mechanisms to enable him to do this. Otherwise, he would not be able to find food, and he might die, which cannot be the goal of any organism.

In that context, it might also be good to note that horses who have done tracking four or five times do not forget this skill, even if there are 6 months in between sessions. They do not need repetition of the basics but automatically return to the level they had before. This shows how important exploration and tracking is for horses. A few short learning moments, without repetition, are enough to store the learning mechanism in the horse's long-term memory. This happens with things that are extremely important to the horse, such as remembering how the lion behaved before he attacked. New experiences that are not as essential are normally first stored in the working memory and are given a place in the long-term memory only through lots of repetition.

2.4 A CLOSER LOOK AT TENSION

The points above bring me to an interesting question: How much tension can a horse have during a training? The study 'Improving the Recognition of Equine Affective States' by Bell, Rogers, Taylor, and Busby looked at how horse people viewed training. It found that many of them failed to recognize behaviour associated with negative affective states and the signals of pain and/or fear associated with it. Of the small group that did recognize these signals, a percentage chose to keep training or having their horses trained this way.[15] This is something I recognise from what participants in my lectures and workshops say: that tension is a part of learning; therefore, I think that it might be good to take a closer look at what tension is.

GOOD STRESS VERSUS BAD STRESS

I think this book makes clear that chronic stress or frequent moments of acute stress have a negative impact on your horse's wellbeing and learning ability. On the other hand, seeking out and experiencing new or familiar stimuli, possibly with some tension, is also a part of indispensable joy in life. How do you translate this into a practical working method? To me, the crucial criterion in distinguishing good tension from bad tension is whether the horse wants to approach the task and willingly engages with it, or if he would prefer not to do it and wants to walk away halfway through (horses will indicate both choices to varying degrees by means of big or small signals). The signals are a bit more challenging to observe with horses who are not used to having the freedom to make choices, such as approaching or walking away without negative consequences being attached to them. In that case, you should watch patiently, making careful note of the subtle signals these horses give. Once they learn that they are allowed to make choices without experiencing negative consequences as a result, their body language will become more pronounced.

Looking forward to something, wanting to experience something, and enjoying something elevate tension. When I do an enriched environment or tracking with a horse, after he has done it few times before, he is often excited. He cannot wait to begin. When horses are tense because they really want to do something, they understand what is about to happen, and they really want to get started. (Note: This presents differently than a horse who is acting out of obedience and would not do something of his own accord.) They walk or trot straight up to the Enriched-Environment objects or the smeller (the starting point) in the case of scent tracking. If I were to take them away before they were done with the activity, they would follow slowly, perhaps glancing back often.

If a horse shows ambivalent behaviour for a time, displaying both a tendency to approach the object and to move away from it, approaching will eventually win out. And even if the horse is not yet approaching the objects in the enriched environment, he is still acting in response to them through his positioning. He is not running to the exit to find a way out; he is working on his approach. It is just taking him a little longer. In both cases, the emotion that wins out is positive, and his state of mind goes back to relaxed.

However, it is also possible that the horse wants to get away from the situation or stimulus, showing a weakened or strengthened flight response, or that he shuts down and shows no more engagement with the handler or the job he is asked to do. These situations are cause for me to stop and change things up, to find a method that will inspire engagement in the horse (in addition to checking and investigating other things, of course).

What explains the difference in terms of biology? It is the involvement of the amygdala, the emotion and fear centre of the body, as the tension is experienced. In his book *Behave: The Biology of Humans at Our Best and Worst*, Robert Sapolsky explained it like this:

When a rat secretes tons of glucocorticoids [which include cortisol, RD] because it's terrified, dendrites atrophy in the hippocampus. However, if it secretes the same amount by voluntarily running on a running wheel, dendrites expand. Whether the amygdala is also activated seems to determine whether the hippocampus interprets the glucocorticoids as good or bad stress.[1]

So here too, activities that do not activate the amygdala (those that do not inspire any fear), which include tracking and enriched environments, have positive effects. Exploration that includes fear, meaning the activation of the amygdala, has negative effects.

> The hormone cycle also influences the way the hippocampus functions. 'Remarkably, the size of neurons' dendritic trees in the hippocampus expands and contracts like an accordion throughout a female rat's ovulatory cycle, with the size (and her cognitive skills) peaking when estrogen peaks'.[1] This is also something we have to keep in mind when living and working with our mares.

TRACKING FROM A CALM AND PROACTIVE INTERNAL MOTIVATION

The example above demonstrates how important it is to avoid fear in the horse, however mild. I have discovered that the way to do this is to have the horse explore or track with a calm and engaged mindset (if necessary, we can help him to achieve this). It is also of crucial importance that the horse maintains control and freedom of choice (after all, the loss of control and unpredictability create tension). That is what this method is based on: carefully creating a programme and adapting it to the horse's ability. That is why reading his body features and behaviours is so crucial. This will come up again in the following chapters.

FIRST EXPLORATION/TRACKING AND ONLY THEN REGULAR TRAINING

> *Prepare the brain before you train.*
>
> ***Dr. Amber Batson, Understand Animals, UK***

In describing all the advantages of exploration and tracking and thinking about their implications, my thoughts automatically also go to the horses who have trouble dealing with our human world or meeting our demands. Luckily, many of them get a chance to learn this and are sheltered for a time in rescue centres or helped by trainers. There is often a lot of pressure on the time set aside for the training of a horse. Time is money, and there are not always enough financial resources to pay for a lot of training time. This often means that, at their rescue centre, horses are trained from the get-go to perform human-centred exercises like lifting a foot on command, walking docilely on a rope, standing still when commanded, not bucking while being mounted, allowing themselves to be touched everywhere, etc.

I can imagine that a large percentage of horses who arrive at rescue centres or training locations have just gone through a period of tension and that this tension probably continues until they have gotten used to their new, temporary accommodation. In the foregoing pages, we have seen that this can cause the horse to have more trouble learning and remembering things we are trying to teach him. It makes a lot more sense, therefore, to offer a horse an exploration and scentwork programme before beginning human-oriented exercises. That way, all the benefits I have listed before are within the reach of the horse and his handler/rider, which is decidedly in his interest, as well as in ours. (About the time aspect: I hope and expect that stimulating this healthier homeostasis in the horse means he will need less repetition when doing the human-centred exercises.)

It is important to note that tracking and exploration are never a replacement for a consultation with a vet or behaviour therapist. If your horse has physical or mental problems, a visit from them is a first indispensable step.

REFERENCES

1. Sapolsky R.M. (2017) *Behave: The biology of humans at our best and worst.* Penguin Random House.
2. Wikipedia De Vrije Encyclopedie (2020). *Gliacel.* https://nl.wikipedia.org/wiki/Gliacel
3. Leblanc M.-A. (2013) *The mind of the horse: An introduction to equine cognition.* Editions Belin.
4. Gadbois S., Reeve C. (2014) *Canine olfaction: Scent, sign, and situation.* https://www.researchgate.net/publication/280446218_Canine_Olfaction_Scent_Sign_and_Situation
5. Panksepp J. Biven L. (2012) *The archaeology of mind: Neuroevolutionary origins of human emotions.* W.W. Norton & Company.
6. Rogers S., et al. (2018) *Equine behaviour in mind-applying behavioural science to the way we keep, work and care for horses.* 5 M Publishing, Sheffield.
7. National Scientific Council on the Developing Child. (2014). *Excessive stress disrupts the architecture of the developing brain: Working paper 3.* https://developingchild.harvard.edu/wp-content/uploads/2005/05/Stress_Disrupts_Architecture_Developing_Brain-1.pdf
8. McGreevy P., McLean A. (2010) *Equitation science.* Wiley-Blackwell.
9. Nyberg J. (2014) 'The adaptable brain', Lecture at International Dog Symposium, Oslo.
10. Cooper J., McGreevy, P. (2003) Stereotypic behaviour in the stabled horse: Causes, effects and prevention without compromising horse welfare. In *The Welfare of Horses,* ed. N. Waran. Kluwer Academic Publishers, 99–124.
11. Sapolsky R.M. (2004) *Why zebras don't get ulcers: The acclaimed guide to stress, stress-related diseases, and coping.* St Martin's Griffin, third and revised edition.
12. McEwen B.S. (2004) Protection and damage from acute and chronic stress. Allostasis and allostatic overload and relevance to the pathophysiology of psychiatric disorders. *Ann NY Acad Sci* 1032:1–7. New York Academy of Sciences. doi: 10.1196/annals.1314.001.

13. Purves D., Augustine G., et al. (2012) *Neuroscience,* International Fifth Edition. Oxford University Press, New York.
14. Kensaku M., et al. (2014) *The olfactory system, from odor molecules to motivational behaviours.* Springer Science+Business Media, Tokyo Japan.
15. Bell C., Rogers S., Taylor J., Busby D. (2019) *Improving the recognition of equine affective states.* https://www.ncbi.nlm.nih.gov/pmc/articles/PMC6941154/

3 *How to implement exploration in daily life*

LET US GO BACK to the practical side. Now that you have read how many benefits there are to be had, you are probably dying to get started. I will deal with the following exploration and empowerment exercises: creating enriched environments, giving choices, and doing scentwork.

3.1 ENRICHED ENVIRONMENTS

An enriched environment is, as the term says, the enrichment of an environment with objects. In my method, this is temporary. You can build an enriched environment in large and small spaces, such as pastures, arenas, and paddocks, and you can also do it in more public spaces.

Depending on the space, the horse, and his development, I choose a number of suitable objects to use, and I put them away after I have used them. In other words, I don't buy a ball, for example, and leave it hanging in my horse's stall or lying in his pasture.

The objects I choose are from real life because those are the things to which you want to accustom your horse. They are things that could be found in a house, shed, or garden, such as blankets, pieces of cloth and towels, buckets in all shapes and sizes, mats in all shapes and sizes, bags in all shapes and sizes, a broom, a small tent, garlands, balls, garden hoses, plastic bottles, trash bags filled with plastic trash or household waste, and so on.

Of course, it is important that the objects you use are safe for your horse so that he can investigate them without risk. This means you need to check beforehand if your horse might choke on something or get his foot or nose caught in it. You can cut the handles on bags, remove caps from bottles, or remove other small, loose items from objects. I also look for a location in which the horse feels safe, or at least safe enough to go exploring. Additionally, the objects have to be able to withstand the horse's drive to explore. He has to be able to take objects in his mouth, stand on them (if they induce him to), smell them, and so on, so it should be okay if objects get dirty.

When I am choosing my first objects, I try to gauge which ones the horse will want to investigate. (If, as an owner, you already know or can guess what your horse likes, or which objects he will be interested in, you have a great starting position.) I also try to pick items that will cause the horse to experience little to no tension from which he is able to recover in that same session. The goal is to begin and end in a relaxed state. I place objects in such a way that the horse is not forced to confront them. I place them some distance away from the entrance through which he will enter the arena so that he has a choice about approaching them or not and can do so at his own pace. Some horses will trot toward the objects with glee, while others will be more timid and feel out the situation for a while first.

Fig. 3.1

I usually begin with three or four objects (**Fig. 3.1**). These will be the core objects. One of these is a rough black cloth of 3 to 4 metres that has little holes in it. In each session, I replace one or two of the core objects with new ones. I leave at least two of the core objects from the previous session so the horse can return to them, recognize their scents, and possibly regain some of his confidence: 'Oh yeah, I know this thing'. It can be that once your horse learns how to handle objects, he is ready to experience a greater number of objects at once, if the space permits this. In that way, your collection of core objects will grow over time.

The decision is always up to the horse. He can decide whether he wants to explore and, if so, for how long. The horse if not forced to do anything. He also determines how often he wants to return to the objects in order to experience them again. I also make sure the horse can easily withdraw from the objects, ensuring that he has an easy 'escape route' if he suddenly wants to leave the enriched environment.

Exploration is thirsty work, so make sure your horse is able to drink before and after the enriched environment. If needed, put out a bucket of water so that he can approach it if he wants to. Eating grass and hay has a calming effect. If there is no grass in the space where you have prepared the enriched environment, you can put a small amount of hay inside it, in case your horse wants to calm himself down. If your horse is hungry, he might start eating it sooner, but that is fine. His curiosity and thirst for exploration will surface eventually, before or after he eats.

WHEN TO USE FOOD

Fig. 3.2

Let your horse determine the learning process. Give him room to experience. Do not use the enriched-environment objects in obedience games, or have your horse touch them on command. That is a fundamentally different process. We humans are no different in this.

(**Fig. 3.2**). Think about finding out how a fidget cube works: you learn which button does what, which button is loose and which is tight through personal experience. That it is very different from touching things on command so you can get a treat. The exploration itself is where the fun lies. For instance, when you are going around a store, and you see a coffee cup made of some sort of unusual, rubbery material, your curiosity is aroused. You lift the cup and touch the material, and then you put it back on the shelf. We do not need to be rewarded for this: the process of exploration was reward enough in and of itself. It answered our internal question: 'what might that feel like?' That is why you do not need to reward a horse with a 'good boy' and a piece of food after he has investigated an object. We are talking about a horse who is already in exploration mode. If you have a horse who does not yet want to investigate the objects, then you can adapt your enriched environment in a way that stimulates him to start doing this. For a very cautious horse, for instance, it would be good to start by only laying down towels. In the very early stages, you can make the towels more accessible by ensuring that they do not have too strong a scent. Because your horse may have negative associations with certain scents, I would use a towel that has been at the stable, not one, for instance, that smells strongly of dog or cat, or one that has spent time in a veterinarian's clinic, as such items are counterproductive to arousing the urge to explore. So, do I never use food in an enriched environment? Sometimes. It is a personal choice that depends on what the horse and owner need, in combination with the specific learning goals. Generally speaking, adding food can lead to different situations: on the one hand, adding food can provide a pick-me-up. Then, after two sessions, you stop using the food, but the urge to explore stays stimulated, and the horse continues to investigate the different objects in the following sessions that do not contain food, perhaps with an even greater interest because searching for food is no longer his first priority. On the other hand, it can also cause a situation in which he stops investigating the objects as soon as he stops finding treats. Or, in a more positive turn, the horse can start to associate the objects with the treats. In that case, he keeps searching for food while looking at the objects, and these things together constitute a positive experience. As such, deciding to add food or not is your decision, also depending on the horse, of course, as well as a possible owner, the circumstances, and the learning goal. Personally, I tend to separate things. I do not use food in an enriched environment so that the emphasis is on discovering the objects. However, in enriched environments with a puzzle and scent tracking, I do use food. In that sense, every variety of exploration and tracking has its own character you emphasize.

If a horse perceives an object that is not familiar to him, it is possible that it takes him much longer to approach it than the time we humans would have taken to do so. Horses and people differ in this. People are accustomed to responding rationally to stimuli, firstly because we know these stimuli (they are from our world, after all), and secondly because making connections between stimuli comes naturally to us. As such, we often think that a horse ought to understand that an object is not scary. After all, 'it is only a…'

We also have a lot of confidence in our own abilities. We are used to stimuli from the human world. The resulting assumption that 'it is probably nothing' is not something a horse can afford. There could be a lion behind that trash can. By practicing caution, or possibly even fleeing, the horse is able to save himself from a lion or some other stimuli. The fact is that many horses take much longer than humans to investigate an object, calm down, and conclude that the stimulus is alright. This can also be true of horses who have been trained, but who were taught in such a rushed fashion that they only half processed certain stimuli, or not at all, causing them to create a negative feeling in association with them, or of horses who were punished in the past when their attention was diverted to the stimulus.

A RECURRING PATTERN

If horses are cautious (without their flight mechanism being triggered in any way), you will often see the following pattern of exploration:

1. They turn their hindquarters to the object (calming signal) while grazing (possibly a calming signal), or lowering their noses to the ground. They occasionally glance backwards when doing this.
2. They then turn their flank toward the object (calming signal), if possible while grazing.
3. They walk past the object in an arc (calming signal) or directly past it.
4. They can, but do not have to, show some other calming signals here, such as blinking, chewing, a head turn, or a seesaw lowering.
5. They can repeat these four steps, sometimes while moving gradually closer to the object, until they think the time is right to make the approach.

It makes sense that a cautious horse would show calming signals when approaching an object. After all, calming signals are signals of appeasement. They are intended to prevent conflict and maintain social relationships. Horses can give these signals to other horses, animals, and people, but also to all sorts of stimuli. (My book *Language Signs and Calming Signals of Horses* contains a lot more information about calming signals and other signals a horse can send, including the body features that go with them.)

Recognizing the pattern of Steps 1–5 is valuable. What might seem meaningless at first suddenly acquires meaning within your horse's exploration development. The different steps vary in the amount of time they take, which means you can sometimes be waiting as long as 20 minutes before your horse is ready to investigate the object, so patience is necessary. Although it is true that if you do more enriched environments, your horse will need less and less preparation time before approaching an object. After a couple of times, he might even be trotting straight toward objects right away.

Picture series: *A few days before this session, we placed a couple of bottles inside the cement tub and scattered food in with them. Now, we have placed the same bottles with some food on the black cloth, also adding a few objects that Indy is familiar with. We wait to see what she will do* (**Fig. 3.3–Fig. 3.12**).

Figs. 3.3–3.12 When Indy enters the paddock, she walks a small circle around the objects, then stands with her hindquarters facing the objects, a pose in which she can still see them well (Fig. 3.3). After that, she approaches the objects, but veers off a few metres away from them (Figs. 3.4–3.6).

Figs. 3.3–3.12 (Continued) She walks to the edge of the arena and starts grazing, all the while keeping her hindquarters turned toward the objects. A little later, she turns her flank toward the objects (Fig. 3.7–3.8). The cycle repeats itself: Indy walks toward the objects (she gets a little closer than last time), veers off, and starts grazing at the edge of the arena with her hindquarters and later her flank turned toward them (not included in the series). After grazing while occasionally glancing at the objects for a time, she walks around them in a circle (Fig. 3.9), then approaches the cloth in a small arc. She sniffs the cloth (Fig. 3.10) and starts to search for the food in the between the bottles.

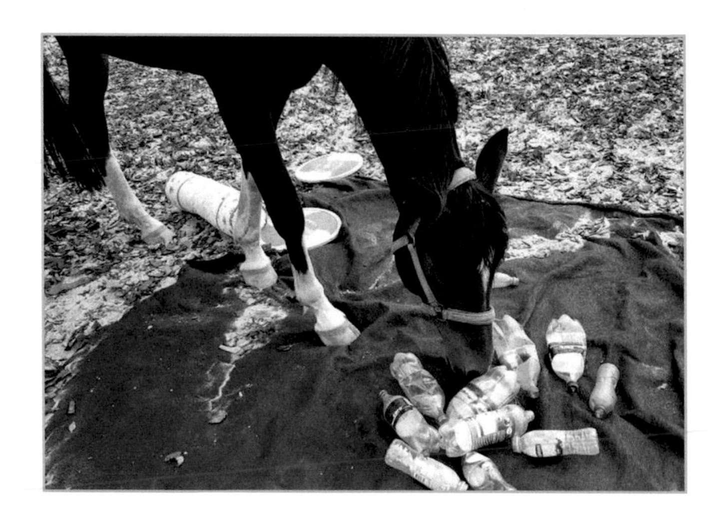

Figs. 3.3–3.12 (Continued) While doing this, she jostles the bottles (Fig. 3.11). She also has no problem stepping across and in between the bottles (Fig. 3.12). It took Indy just over 4 minutes to touch the bottles. Next steps are putting the bottles on the ground and then putting them in different locations. That way, we practice with the bottles Annemarie might encounter along the side of the road while riding in the country. Indy might see them, and they might make noise.

AN ENRICHED ENVIRONMENT WITH SCENT

You can also choose to enhance an enriched environment even further by adding scents. I often use towels for this. I place a scent on each towel for the horses to discover. This is a spin on the scent gardens for dogs introduced by Anne Lill Kvam.

We do not naturally tend to think of socialising our horses to scents, but this exercise is especially good for that. And consider how many smells our horses encounter when they are in the country, at a competition, or at the vet's. A few come to my mind: the pesticides farmers use to spray their fields, the personal odours of passing hikers, the smell of wet dogs, and so on. At competitions, your horse might smell chips, hamburgers, and popcorn. At the vet's: the scent of disinfectant.

For safety reasons, do not practice with pesticides, but you can work with many other scents. You can place towels in a dog bed and, after a few days, put them into your horse's enriched environment. You can ask friends to keep a towel in their living room, so that when you pick it up, it will smell like their living space. If your vet is willing to do the same, it is good to do this there, too. Put the towel in his practice's waiting room for a few hours and then pick it up (make sure the towel was in a safe place though: it should not be infected with the pathogens of other animals). I can place (or eat) a hamburger and some chips outside of the enriched environment, on the other side of the fence, so that the scent reaches the horse on the wind. You can also bring some scents from home: You can drip some pickle juice onto a towel, or some tea, or a drop of soy sauce, diluted with water. Remember that scents we can smell well are especially pungent to the horse, so do not choose scents based on what you can smell, but tone it down considerably to ensure that your horse wants to investigate the scents and is not overwhelmed by them.

TOWELS AS AN AID

If your horse is being moved to a different stall or facility or he has to go to the vet, and you have some time to prepare for this, you can place a couple of towels in the aisle of your horse's stall or hang them up on a rack. You can weave the towel through the bars of the stall of his best horse buddy, too, if this horse does not destroy things like that. You save these towels in a closed box, and then you can hang them up or weave them through the bars of your horse's new stall or at the veterinary hospital. That way, your horse can stand by the towels and smell the scent of his home or his horse friends. If you want to do this, do not use towels that carry a strong scent of detergent, or it will overwhelm the familiar scents of the stable.

I have placed the following scents on towels: tea leaves, three drops of lavender, and a small dollop of ketchup (there is no food present) (**Fig. 3.13–Fig. 3.25**).

Figs 3.13–3.25 Hope enters the arena and trots around the towels. She then walks toward the first towel in an arc. She does not give herself time, instead sniffing it for only a fraction of a second before moving on to the next towel (Fig. 3.14 and Fig. 3.15). This second towel has a fresh, smeared-out ketchup stain on it. She sniffs at the towel while walking over it, turns to the right, and trots away (Fig. 3.16).

Figs 3.13–3.25 (Continued) She trots slowly across the first towel (Fig. 3.17), sniffing along the way. Then she trots in a turn to the left, ending up at the first towel once again (Fig. 3.18) and trotting over it. She trots in a short turn to the right and slows to a walk. She steps over the first towel, sniffing it as she goes (Fig. 3.19), does the same thing to the second towel (Fig. 3.20).

Figs 3.13–3.25 (Continued) She then walks in a circle around the paddock. After that, she starts to graze at the edge of the paddock with her flank turned toward both towels, but her eyes are on the one on the right (Fig. 3.21). And indeed, there she goes (Fig. 3.22). She walks across it while sniffing it. After that, she stops next to the towel on the left a couple of times to smell it while standing still (Fig. 3.23–3.24). She then moves to the exit and indicates that she wants to leave the arena. The entire sessions lasts no more than 10 minutes.

Fig. 3.25 Vos gives a very different reaction to the ketchup: yum, tasty!

AN ENRICHED ENVIRONMENT WITH A PUZZLE OR SOUND

Once your horse has discovered how much fun it is to discover things, I often introduce a new series of different enriched environments. These have an object in them that challenges the horse to be proactive and brave. It also appeals to the horse's problem-solving ability. Sometimes I also plan it in such a way that the activity releases sounds. Sound is a stimulus that all horses encounter; if you put an object into the enriched environment that produces sound, the horse is the one who can influence the start and intensity of the sound. Also, because the sound comes from the objects, there is a certain logic to it. 'Ah, that's what it sounds like when I knock over two plastic crates. And this is what it sounds like when I knock over three plastic crates full of cans. This is what it sounds like when I push a full trash bag, or when a rubber object falls onto stone'.

For these types of challenges, I do use food, in which case, getting the food is the reward the horse will get for participating in the activities. I make a list of the top 10 rewards so that I can make a conscious choice about which reward or rewards I will use.

You begin with the activity you think your horse can handle. If he does, you can gradually add more challenges. Because of the earlier activities, he has had a chance to build up his problem-solving ability, so now he has a greater arsenal of skills at his disposal. He will also have practiced holding his concentration, and possibly a particular body posture, for longer. The challenging activity is one of the enriched environment objects that is set between other objects with which the horse is already familiar. For instance, you put some food beneath objects the horse can shove aside or lift, or you place food inside rolled-up objects. You can also put food inside a crate with a loose lid on it so that the horse can slide the lid aside, causing it to fall to the ground, and get at the food. Another option is placing food inside a tub and then placing a towel over it. This can also be a wet towel, which is different again in terms of weight and texture (**Fig. 3.26– Fig. 3.28**).

Figs. 3.26–3.28 Indy enters an enriched environment. She knows all the objects on the floor: the cloth, the objects, the cement tub. We have put some food into the tub. She is used to eating from this. The only thing that is new is that, behind the cement tub, we have hung up a cloth that flaps in the wind a bit. Indy approaches the cement tub and takes a few seconds to sniff it (Fig. 3.26).

It is important to note that there is a difference between solving a puzzle and then repeatedly doing this puzzle again. When you are appealing to your horse's problem-solving ability, such that he is motivated to solve a puzzle himself, you are challenging him mentally. However, a discovery can only be made once. Once the horse has solved the puzzle, you are no longer mentally challenging him, but instead giving him a game he can repeat, which of course can certainly be a fun activity. Looking at this from within the biological framework I have shared before, you can say that finding the solution could have established a new neural link in the brain. The times after that do not create new links, but they do strengthen the old link. Additional myelination has taken place. If you change the game, then you might once again surprise the horse, spurring the production of new brain cells in the memory and the creation of new links.

Figs. 3.26–3.28 (Continued) She then begins to eat, certainly not at full relaxation. Her hind legs are not in a relaxed pose directly underneath her body (Fig. 3.27). Her legs are in a pose that is a bit more relaxed in Fig. 3.28. However, more relaxation is certainly possible. This does not happen during this session.

TOGETHER OR ALONE?

A lot of people ask me: 'Should I let my horse discover on his own, or can his buddy also participate?' This is a difficult question. The answer depends on different factors: the horses in question, the space in which you do it, and the learning objective. To me, the goal of an enriched environment is to empower the horse. Because horses are herd animals, they can experience so much more stress when they are without the members of their group, so individual practice can help him to learn that he does not need to feel this stress and that he can handle things on his own. That is why I prefer to practice alone with the horse. If, however, you think it will help to let the horse have his buddy with him the first few times he does an enriched environment, you should do it. There is no one size fits all. See what might help your individual horse. Maybe you could have them discover the enriched environment together and then have the horse you meant to practice with do it by himself. That way, the horse has already learned how much fun it is to discover objects, and he is inclined to start doing it on his own.

If you have several horses and not enough time to challenge them all individually, you can also place multiple objects in the pasture for them to discover at the same time. Just like eating and sleeping together, discovering together is also a natural social activity for horses. And you also capture a lot of previously discussed benefits with this. I am assuming here that your group of horses are well suited to each other and that there is not one among them who is prone to resource guarding (**Figs. 3.29–3.32**).

Figs. 3.29–3.32 Tracie Faa-Thompson's horses are discovering a rug together.

When I construct an enriched environment, I see myself as a director creating a perfect stage setting for my horse. I want to set it up in such a way that the horse starts out relaxed and ends relaxed. If the horse feels tension, it is light tension from which he can recover on his own. Wherever possible, my role is to observe and to gauge which objects interest my horse and which do not. I also try to answer other questions, which I discuss in section 4.1: 'The enriched environment as an assessment tool'.

I think it is very important that the horse overcomes his own tension barriers. If he cannot do this, then I will build another enriched environment in which it is possible, rather than trying to help the horse on the spot. If I do help (this desire is often expressed by the owner), I or the owner will stand next to some of the objects, if this reassures the horse. Sometimes I also help horses whose problem is not that they are scared of the objects but rather have been conditioned not to touch anything; I encourage them to do so anyway. Sometimes I only need to do this once or twice before the horses start discovering on their own.

People are full of plans, goals, and expectations. Sometimes they have built a nice enriched environment, come up with a game, or given their horse a choice, and the horse does not react to it. And they already had such a nice picture in their heads of a horse having fun, playing, and discovering. The tendency to want to help the horse is strong. Do not give in to it. Remember that this is about the horse's learning process, about his choices, his freedom. He can choose not to act, if he wants. And what we consider unsuccessful or incorrect might not be at all.

If you have built an enriched environment, and your horse does not explore it all, look to what he has shown. This is so enriching and tells you so much about your horse. Film him, get a timer, bring a notepad and jot down the body features and behaviours he shows in response to the enriched environment. Maybe your horse needs more time than you thought, is not feeling well, is making progress in much smaller steps than you had envisioned, or prefers to discover different objects.

UNFENCED AND REAL-LIFE ENRICHED ENVIRONMENTS

Figs. 3.33–3.35 Joy and Suzy are out for a walk. Joy is choosing where they go.

Figs. 3.33–3.35 (Continued) When she sees a new object, in this case a roll of wire netting, she is given all the room she needs to investigate it in her own way, meaning Joy chooses the distance she wants to maintain from the roll and also when she chooses to leave.

Once the horse has discovered a number of enriched environments in a fenced area and can do so calmly, you can also start building them in unfenced areas, or take him for walks or rides in places where he can find natural objects, allowing him to discover them in a so-called real-life enriched environment (**Fig. 3.33**–**Fig. 3.35**). With both forms, I make sure I impede the horse as little as possible in his investigations. That way, the horse can move his head freely and have a good look at all the objects at various heights.

Important: Do not cut the horse's whiskers, or vibrissae. These thicker, longer hairs sticking out around his nose, mouth, chin, and eyes function as tactile hairs that provide the horse with special information when he is approaching an object that help him to avoid bumping into things (**Fig. 3.36**).

Fig. 3.36

3.2 CHOICES

Nothing is more empowering for your horse than letting him make his own choices. Having self-determination over your life, or elements of it, is so beneficial. It is the polar opposite of the learned helplessness and depression that stems from not being able to influence your surroundings. Making choices stimulates feelings of joy. Not only that, but the experience and ability to influence your environment causes a reduction in tension. I am convinced this is true of all mammals. Of course, ensuring the safety of you and your horse is a precondition to freedom of choice. This must always be the underlying foundation.

The moment you let your horse make a choice, it has the following benefits:

- You can see in a very direct way what your horse wants and does not want, what he does and does not like in that moment.
- Making a choice forces your horse to consider, weigh his options, and proactively follow through on his choice.

The benefits discussed in section 1.3 also apply when you stimulate your horse to make a choice. However, the following conditions do apply:

- The choices have to be real judgments for the horse, and they must have value to him.
- The choices should not be forced by you, the handler, putting pressure on the horse. You do not want to activate the horse's fear system, either strongly or weakly.

- The choice he makes can in no way (active/passive/verbally/nonverbally) be punished. If this has happened in the past, you will sometimes see horses who joylessly conform to their owners' desires because they are afraid to do anything else.
- Your horse's choice is always okay and is always acted on. Only then can choice have true value.
- A choice without a real alternative is not a true choice. A horse standing in a sandy paddock by himself will choose to do training more quickly than a horse who is in a verdant pasture with his buddies.

Here are some choices you can think of. These are also mentioned in my book *Language Signs and Calming Signals of Horses*:

- The horse chooses the direction the two of you go in when walking or riding in the country.
- The horse chooses the pasture in which he wants to be released.
- The horse gets to choose his own food rewards.
- The horse gets to choose how he wants to position himself while he is being groomed and saddled (**Fig. 3.37**).
- The horse gets to choose on what spots on his body, where at the facility, and for how long he is groomed.
- The horse gets to choose when he wants to take a break during grooming or some physical treatment.
- The horse gets to choose whether or not to approach a stimulus.
- The horse gets to choose which other horses to greet.
- The horse gets to stop and look around during functional walks or rides.
- The horse gets to choose not to participate in activities that have been devised, even if the activity is making a choice.

Fig. 3.37 Vos (on the right) and Patricia (on the left) are in the grooming area (not in their stalls). Normally, the horses face the other way and are secured in the cross ties. However, we leave them untied during grooming sessions, allowing them to turn if they wish. This is an example of how you can give your horse some freedom of choice in a small way.

Another form of freedom of choice is this (**Fig 3.38**):

Fig. 3.38 Tracie's horses' hooves are being trimmed. The horses are not tied up for this.

During Turid Rugaas's International Dog Trainer Education, we had to take our dogs for a walk. The dog was allowed to decide how long the walk would last and where we would go. Everyone was curious what would happen, the general expectation being that we would never find our way home again. I thought the same. Just in case, I had asked my mother to come pick me up from wherever I ended up. So, off I went. Turns out, after a beautiful two-hour and twenty-minute walk, I was back at my front door. My course mates fared the same. All of them ended up back home after two-to-three-hour walks. This exercise inspired me to do the same thing with my horse. I walked him to a good spot a 100 metres away from his stall and pasture buddies. He could take one of two paths into the forest, turn and go back home to his friends, or he could stay there and graze in the spot where we were. I was so curious to see what adventure we would have together. Unfortunately, one half hour of grazing was followed by another half hour of grazing, and then another, and another... After three hours, I was at the end of my tether. Yes, I had been patient and learned a lot about my horse, but this had not been an altogether inspiring episode. From that moment on, I began to introduce two hand signals.

TWO HAND SIGNALS: 'YOUR TURN' AND 'MY TURN'

These two hand signals symbolize when you give freedom and when this freedom stops. For instance, if I am going on a walk, I can give the 'your turn' signal, literally taking a step back at the same time. The ball is in Vos's court then. If the need arises, or I am ready to resume the lead, I make the 'my turn' signal, and the initiative will revert back to me (**Figs. 3.39–3.40**).

Fig. 3.39 I give the hand signal that tells Vos he can take the lead. I also take a step back. Vos chooses to take a walk, during which he chooses to graze every once in a while.

Fig. 3.40 I give the hand signal that tells Vos I am taking back the lead. This means that, on this walk, he is no longer allowed to graze, but he can decide where he wants to walk (in front of me, beside me, behind me, etc.).

This role reversal is emphasized more with the hand signal. Oftentimes, you will not need it, because the reversal is automatic. When you are grooming him, your horse can move a few paces in order to indicate where he wants to be brushed. Then he can step back, or you will move on. When you give your horse more freedom, habits, behaviours, and emotions change. At first, this can take some fine-tuning. Some horses flourish when they get more freedom and stay very manageable. Other horses flourish but also want freedom at times when it is not always possible. Some horses do not know what choices they want to make. Start off slow, and see what works for both you and your horse and what is possible at the facility where your horse is housed and its surroundings. Wait and see what happens, and then build on that.

Situational leadership and reciprocity are strong building blocks for a relationship, both of which have to do with seeing one another's nature, meeting each other halfway, and wanting the other to be happy. And the special thing about reciprocity is that you will see that the more you let your horse have what he wants, the more he is willing to do something for you in return and conform to your wishes.

3.3 WHEN DOES IT TAKE LONGER TO SEE RESULTS?

Exploration, tracking, giving choices and the many benefits they have for the horse's wellbeing and his ability to live and work with us have been explained in previous chapters. It is also important, however, to tell you about the pitfalls you might encounter. There are even situations that might delay or even entirely nullify the many benefits. When do you see less or no result, or when does it take longer?

- When you do not give your horse enough freedom to discover and are too controlling during the exercises, keeping the focus of the learning experience from moving to the horse but keeping it on his dependence on you and your control of him.
- When your horse's life and training is so controlled and organised that the enrichment or tracking exercises can take place only sporadically. Those few times then cannot compete with the experiences and emotions engendered by all the other times he does things according to a different method, one in which he is always in a dependent and controlled position.
- When your horse's life, independent of the exercises, is such that the positive experiences he has during exploration and tracking (temporarily) cannot compete with his circumstances. This can be the case, for instance, when your horse is in pain; too tired to explore; not in contact with other horses and suffering from it; kept in his stall 24 hours a day and feeling significant tension because of it; or if he does not have a suitable place to sleep, making it impossible for him to achieve REM sleep, which deteriorates his memory. If your horse lives in circumstances like these, it does not mean you should not do the suggested exercises. It is just

that his circumstances might mean that it takes longer for you to see results. If you have a horse who is regularly exposed to moments of stress, or who feels chronic stress, I do want to strongly advise that you consult with a veterinarian or a skilled behaviour therapist. That way, you can make a plan in which all the elements of your horse's life are analysed.

- If there is a large discrepancy between the training methodology that is used on the horse and the exploration, freedom of choice, and scent tracking described in this book. For instance, if you are accustomed to physically punishing your horse when he refuses to walk past an object, and you start no longer doing this because you think the horse should be free to investigate objects, the horse might not go ahead and do this right away. The connection the horse might have made between discovering new objects and undergoing the pain of physical punishment will first need to fade.
- When you set stimuli in an enriched environment or real-life enriched environments in such a way that, rather than feeling stimulated to explore and seeing this as a positive thing, the horse's flight response is activated, which may lead to seeing fewer results, no results, or even adverse results. When that happens, he can store the objects in his memory with a negative connotation attached to them and may even begin to generalise this. Of course, this can happen sometimes when you are trying to see what works. It becomes a problem, however, when it happens every time. Then, you are training your horse to become more and more scared, which is why reading your horse is so important.

A SUCCESSFUL SESSION

Navigating and finding out what your horse can do is a journey of exploration in and of itself. To me, an enriched-environment session is a success when the horse, in an engaged way, investigates objects from within his comfort zone and explores objects that are new to him. This horse walks around, investigates, walks away, investigates again, and after some time, he is done.

To me, a session is also a success when the horse is both in his comfort zone and, at times, slightly outside it. That way, you get a kind of back and forth, moving from just outside the comfort zone to back inside it, allowing the horse to recover, go back outside it again, and so on. By operating just outside his comfort zone at times, the horse can practice calming himself down. The stimuli should have been selected in such a way that this horse ends the session within his comfort zone.

If you are in doubt about how your horse is experiencing his enriched environment, give him as much freedom as possible to indicate what he wants. Then, either on the spot or later by means of video footage, determine your horse's engagement and comfort zones using the communication ladders and the information in the Appendix.

4 *How to use exploration to address problems*

IF YOU HAVE a horse who needs more care and attention because, for some reason, he cannot handle his life circumstances and the demands being made of him, exploration and tracking can help in various ways, including:

1. as an assessment tool, possibly in a session with a behaviourist
2. as a tool within a behaviour adjustment plan
3. as a way to bring the horse into balance mentally and physically
4. as a stand-alone exercise

4.1 THE ENRICHED ENVIRONMENT AS AN ASSESSMENT TOOL

I use enriched environments as an assessment tool to help me get a general impression of what a horse is like. You can do this with your own horse or with someone else's. When I am asked to assist with a horse whose owner is experiencing problems with the horse's behaviour, I always bring a number of props that I can put down at the location and use. As such, a first consultation with me consists of a conversation with the owner and then bringing the horse into an enriched environment, after which he is returned to his stall or pasture for a break. I then have another conversation with the owner in which we begin to decide together what the horse needs and how we can devise steps to provide it. I end with another practical part in which we practice the first few steps we have discussed.

I select the location of the enriched environment with care. You want your horse to be comfortable in this place. For horses with separation anxiety, you may need to keep your horse close to other horses, or perhaps bring another horse along and let him graze outside the enriched environment. If these options are impossible, you can choose to have your horse do the enriched environment with a buddy. Later, when the horse has gathered more courage, he can start to do the enriched environment alone.

When I have put the objects down and the horse is in the arena, I like to film him. I then review his body features and signals and try to answer these questions:

- How does the horse react when entering the space?
- Is he comfortable on his own?
- How long does it take before he starts to investigate objects?
- Does he investigate objects?
- If he does not, what does he show, and does it change during the session?
- If he does investigate objects, what senses does he use?

- Does he inspect an object multiple times or only once?
- Does he have a preference for or aversion to certain objects?
- For how long does he discover? If it is only a short time, can I detect the reason? (For instance, does he seem disinterested? Does he know the object? Is he having trouble concentrating or maintaining a physical posture? Is he afraid? Is he being distracted by the actions of others?)
- If he ignores certain objects, does he change his mind within the same session?
- If he shows ambivalent behaviour, such as approaching and retreating, does he do this for a longer time, or does the ambivalence fade away, leaving only one of the behaviours? Which one?
- How is his physical state and how are his movements? Do I see anomalies?
- For how long does he want to stay in the arena?
- Is his behaviour in his stall or at pasture different from what he shows in the enriched environment?

I also look at the owner/handler of the horse during the enriched environment. This makes me understand her better, which helps me to create a plan that is good for the horse and is supported by the handler. To do so, I ask myself these questions:

- What is the handler doing in relation to the horse?
- How difficult does she find it to not influence the horse?
- What does the horse do in relation to the handler?
- Can I discover underlying motivations and emotions?
- What words does the handler use to describe the horse?
- What words does she use to describe the training goal and the enriched environment?
- How does exploration and tracking fit into the facility where the horse is housed?

AN INSTRUMENT TO MEASURE PROGRESS

When you start working with these exploration and tracking exercises and begin to empower your horse and let him make choices, he will get a better understanding of the world around him. This process is not a quick fix that you can get done in a single session. It is a process that influences different underlying emotions and the internal physiological functioning of the horse. Changing these takes time; the amount of time it will take depends on the horse and his physical and mental health at the time, the experiences he has had in his life, his current life, and the demands that are made of him. When I am helping a horse and handler, I will come back several times after a certain amount of time has passed to evaluate and create a follow-up proposal that will take them another step closer to the described goal. During this evaluation, I also create an enriched environment. I film these sessions, too, in order to compare them with previous footage. These visuals give you data that will help you to track your horse's process and chart it so that you are not just dealing with feelings and general ideas. During the follow-up sessions, I also focus on the points of attention I have just listed, and I compare the two videos noting the differences I see. For instance, you could see that, in comparison with the previous enriched environment tape, the horse now no longer shows ambivalent behaviour and starts to proactively investigate straight away. In the case of another horse, you might see fewer tension features and signals or that he is no longer showing displacement behaviour or giving calming signals. You might also see that he is able to concentrate for much longer or that he is able to investigate objects in a different way, for instance, using his tactile sense to a greater degree and manipulating objects more. It is painful to watch those horses who do not start to investigate during their first enriched environment. They seem to lack the internal drive to do so. (It is not the case with these horses that they have learned that investigating objects is not allowed.) These horses do not have the energy to discover. Their lives use it all up. Once these horses get started on a behaviour adjustment plan, meaning they have a bit more energy left, you will see them perk up and start to investigate.

If the handler has built her own enriched environments with her own objects in between consultation sessions, I will sometimes use some of my own objects again during a 'comparison enriched environment'. These will be one to three of the objects the horse also encountered the last time I built an enriched environment for him. It is interesting to see in that case whether the horse remembers the objects. If I put down new objects, I am curious to find out if he will start to generalise yet and connect the new objects to older concepts he has already experienced.

Note: For those who were disappointed to read that this method is not a quick fix for your horse's problematic behaviour, do not feel bad; quick fixes are often forced behavioural changes that have a high rate of recurrence of the problem behaviour. The problem behaviour can also flare up again if the handling of the horse changes just a little or if a different person does it. A quick fix does not address the underlying emotions and physiological processes of the horse. If there is high underlying tension, there is a chance that this will not lessen, increasing the risk of stress-related illnesses. On the other hand, if you do choose to use this longer term methodology, in which emotions and physiological processes change, causing the horse to achieve healthier homeostasis, the changes will be much more durable, and the chances of recurrence are much smaller. You will also empower your horse to become capable, meaning he will also have far better chances to be able to deal with a different handler or slightly different treatment.

4.2 THE ENRICHED ENVIRONMENT AS A TOOL WITHIN A BEHAVIOUR ADJUSTMENT PLAN

As I have said before, I generally alternate between conversations with the owner and practical sessions with the horse during a consultation. In the first conversation, I chart the problem the owner is having with the horse, amongst other things make an inventory of the horse's living environment, and weigh what is possible and what is not possible. In the practical part, the horse can discover a suitable enriched environment. For this, I use the list of questions in section 4.1 for using an enriched environment as an assessment tool. Armed with this knowledge and the information from the first conversation, I create a plan together with the owner that we both find promising and feasible. If the horse can handle it, then we practice the first steps of the exercises that suit the horse. If necessary, we also plan consultations with veterinarians or other specialists. (Sometimes the veterinarian and other specialists visit before the behaviour consultation.) Then there are follow-up consultations, and we talk on the phone or over email where there is a need.

My behaviour adjustment plan contains the following five elements:

1. Drawing up a tension map: The purpose of this is to chart whether there are elements in the horse's life that cause enough tension to negatively impact his health, wellbeing, and learning ability, and if so, which ones. You will want to change these elements to allow the horse to recover and bring him back to a relaxed state. In doing so, you are trying to create a comfort zone for your horse.
2. Preventing the horse's problem from being activated and coming back again and again: In most cases, the problem causes tension in the horse. The goal, however, is to bring the horse back to relaxation in his own comfort zone and to adapt the handler and horse's life so that the tension is not activated or amplified. This is temporary.
3. Stimulating the urge to explore: Together with the owner, I see which exploration or tracking exercises suit the horse and her. We make a start with these exercises. These exploration and tracking exercises help the horse to achieve mental and physical balance more quickly, thereby building a good foundation for learning, remembering, and being together.
4. Doing relationship-sustaining and relationship-building activities:_Together with the owner, I see which relationship-sustaining and relationship-building activities would suit the horse and her. We make a start with these activities.
5. Make a plan to incrementally practice solving the problem that is not being activated at the moment: This is only possible if the horse is mentally and physically ready for this. If he is not in a good place mentally, Elements 1–4 will be executed first. Then, if the horse is relaxed enough and open to taking in and remembering new information, we start on Element 5 (and it does not replace Elements 1–4). Wherever possible, I try to integrate these new habits or behaviours into exploration and tracking exercises. However, I also use habituation, progressive desensitisation, and counter conditioning. In my work, I do not use methods the horse might experience as unpleasant or painful. (In all this, I assume the wishes and demands of the owner are realistic and within the horse's physical and mental abilities).

A CASE STUDY: SEPARATION ANXIETY

I will share an example from the field to illustrate the five points above. I chose a case involving separation anxiety. I will only share the general methodological points.

This case involves Esther and her horse Patricia. Patricia is a Dutch Warmblood mare who is 21 years old. During the day, she is at pasture with other horses, and at night, she has an individual stall in a stable that houses six other horses. Esther rides dressage with Patricia in the stable's arena. Esther would also like to take Patricia on rides and walks in the country, but Patricia does not want to leave the compound, and builds up so much tension on the way that Esther no longer enjoys the walk or ride in the country. The first problem Esther wants to tackle is that she wants to be able to take Patricia out of the pasture in order to groom and saddle her in the stable. So far, she has not been able to do this, because Patricia keeps running out of the stable. We do the following things:

1. Creating a tension map: We assessed the extent to which Patricia's life meets the criteria of the 10 freedoms (see section 1.2). We discussed the activities Esther does with Patricia and which of these Patricia experiences as relaxing or stressful. In the foreseeable future, Esther will only do things with Patricia that do not cause tension. In her case, this means no more longeing. We also removed Patricia from the pasture and took her for a walk around the compound. We watched her body features and signals. We brushed up against the zone Patricia cannot handle because it is too far removed from the other horses, but we did not linger there. Moving from relaxation to high tension, we charted Patricia's green, yellow, orange, and red zones (**Fig. 4.1**).

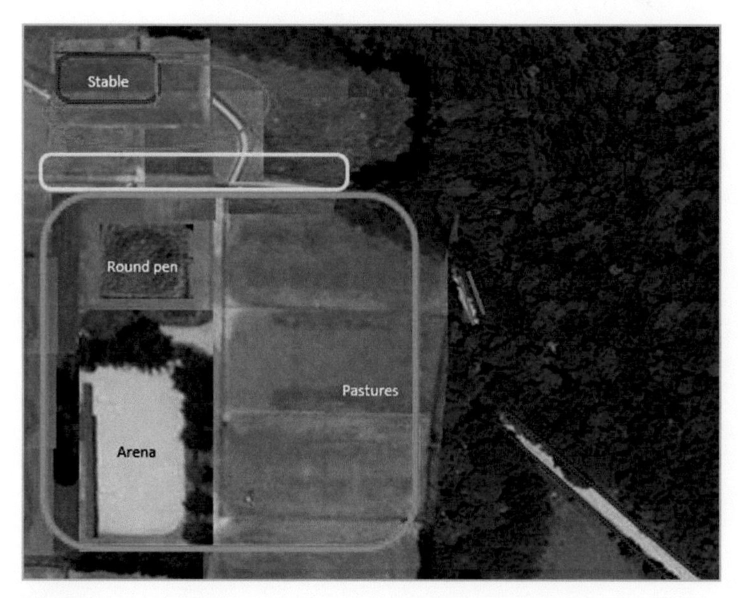

Fig. 4.1

Also, we built an enriched environment during a follow-up appointment two days later (this was to spread out Patricia's activities).

Creating the first enriched environment: (**Figs. 4.2–4.7**).

Fig 4.2–4.7 Patricia investigates the objects that have been placed in the paddock. She starts a pattern of approaching and retreating. She walks toward the objects and then walks away again. As she approaches and retreats, a number of calming signals are visible.

Fig 4.2–4.7 (Continued) It takes her 4 minutes to sniff the first object in the enriched environment, and she sniffs the others right after that. She also walks across the black cloth. After this initial exploration, she releases tension by rolling. I classify this roll as displacement behaviour (she is more rushed, spends little time finding a spot, and briefly rolls on only one side. Her daily rolls in the pasture that are not displacement behaviour look different. For these, she takes her time finding a spot, rolls on both sides, and extensively moves her head in the sand). After that, she returns four more times to investigate the objects. The entire session lasts for 19 minutes.

It was also noticeable that Patricia's eyes were constantly infected, the lymph nodes in her neck were swollen, and her neck and back seemed to be locked up. On the advice of the vet, Esther cleaned Patricia's eyes with water that has been boiled and then cooled down. We also made an appointment with Nanne Boekholt, an osteopath who treated Patricia twice over a period of time.

2. Preventing the problem the horse has from being activated and recurring again and again: We did not want Patricia's separation anxiety to keep coming back. Temporarily, Esther did not work with Patricia in the orange and red zones, meaning that when Esther wanted to groom or saddle her horse, she did this outside in the green zone (at the moment, the stable was still in the orange and red zones). That way, Patricia was able to reach her point of relaxation again more quickly. This also means that Esther temporarily did not take Patricia on rides or walks outside of Patricia's comfort zone.
3. Stimulating her urge to explore: We chose to offer Patricia enriched environments with the goal of doing this twice a week.
4. Doing relationship-sustaining and relationship-building activities: Esther took Patricia on walks twice a week in the green and yellow zones. On these walks, Patricia got to take the lead and choose where she wanted to walk and graze (we wanted to encourage her to build up her self-confidence with this).
5. Make a plan to incrementally practice solving the problem that is not being activated at the moment: Those times that I was with Esther and Patricia, we practiced being around the stable, a bit further away from the other horses. We did this by creating enriched environments at the edges of the yellow and orange zones. We also created an enriched environment inside the stable, adding little piles of hay. That way, when Patricia walked toward the stable and in the stable, she could get food and discover there as well. It is crucial here that the horse can decide to leave the situation at any moment. If she walked away, we followed her.

HOW THE BEHAVIOUR ADJUSTMENT PLAN IS WORKING OUT

Gradually, Patricia became braver and braver. She walked to the stable of her own accord. In the beginning, the smallest sound or scent could have her rushing out of the stable (which we allowed), but as time went on, she scurried around the stable for longer and longer periods of time before leaving. The moment she was relaxed inside the stable, for instance while eating some hay, we started to groom her a little. In follow-up sessions, we sometimes held her in place while grooming when she wanted to start exploring again, for instance in order to put away the brush; after this brief moment, we continued. This method has enabled Esther to groom and saddle Patricia in the stable while Patricia eats hay. It must be noted, however, that Esther gets all her things ready beforehand, holds Patricia on a long rope, and does not tie her up (although she will be able to in the future). Patricia is more relaxed when Esther is present for these practice moments. The relationship-building exercises have contributed to this as has the freedom we gave Patricia to leave the practice site if she wanted to, regardless of whether she pulled on the rope while doing so (at first, Esther walked a bit more slowly as she followed; this happened only twice). Patricia relaxes more easily now, which means she does not walk so quickly anymore.

Picture series over time (**Figs. 4.8–4.22**):
The first time in the stable; Patricia wanted to leave quickly:

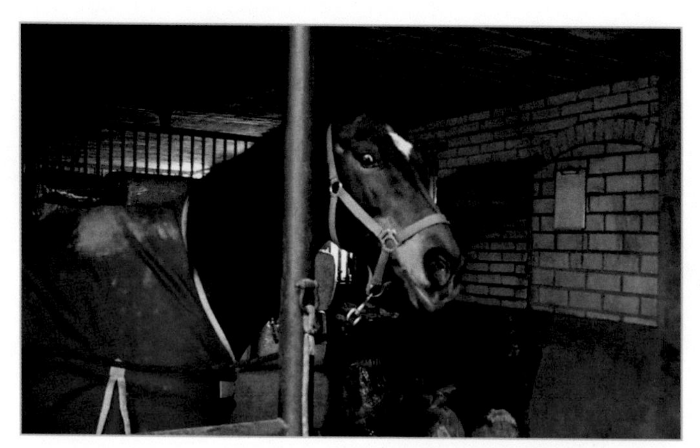

Fig. 4.8 Patricia wants to leave the stable. Her eyes are round, and the whites are showing, even when she is not looking at me. She also has round nostrils, a high head and neck position, and a quick step. The slit of her mouth is elongated. You can see this when a horse is about to or wants to leave a situation. Patricia's lower lip is very slightly tensed. She naturally has a very relaxed lower lip that rests away from the upper lip with some room in between. See Fig. 4.9.

Fig. 4.9 Osteopath Nanne Boekholt at work. After two weeks, Patricia no longer has eye infections. The swelling of her lymph nodes lessens considerably, and her head, neck, and back become more and more flexible.

Figs. 4.10–4.11 Patricia finds the stable increasingly more interesting and enjoyable. She is in a rush to get there from the pasture. In the stable, she is also more relaxed in her body features and signals. She leaves in a relaxed state, too.

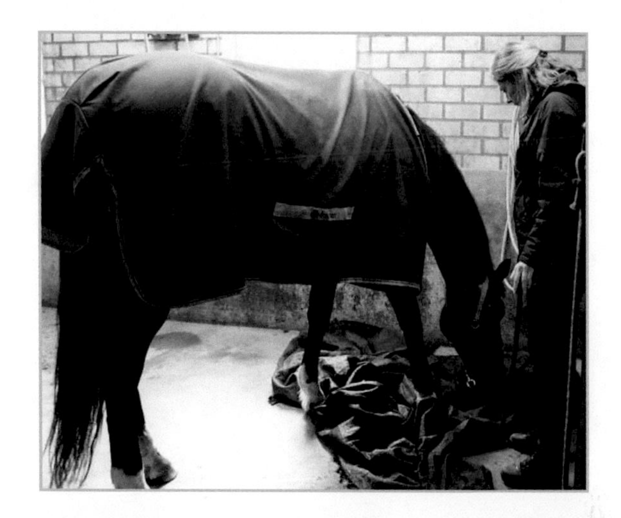

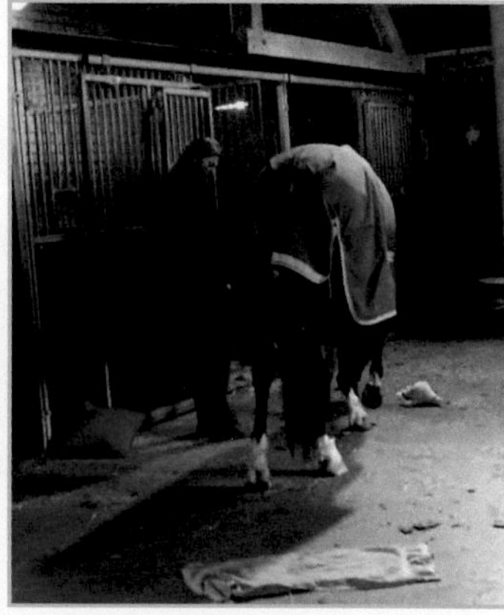

Figs. 4.12–4.15 Examples of enriched environments in the stable. Every time, there is something new to discover, whether it is others' food or stalls or objects that have been placed there to be discovered.

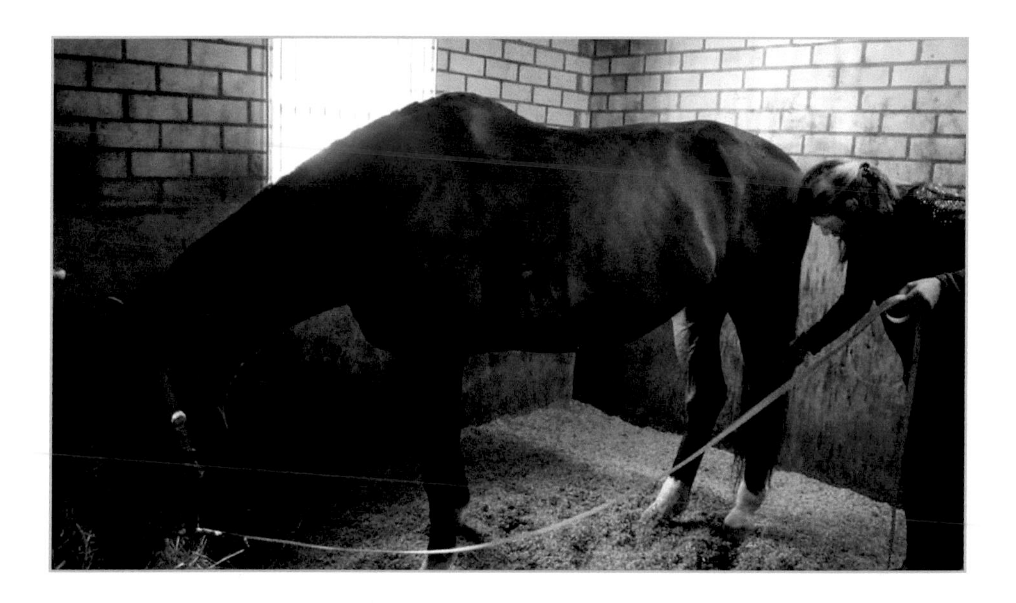

Figs. 4.16–4.17 Esther is able to saddle Patricia when Patricia eats. This can be either in the stall or in the stable's hallway.

Fig. 4.18 During the sessions, the zones shift. After 2 months, Patricia is relaxed in multiple locations in the stable compound or outside it.

Figs. 4.19–4.22 Patricia has become particularly free in enriched environments.

Figs. 4.19–4.22 (Continued) She walks or trots toward new objects, sniffing the first object within 15 seconds, even if these objects are entirely new and unfamiliar to her.

IMPORTANT: DISOBEDIENCE – HURRAY!

When Patricia started to find the world more and more interesting, and she started to feel less tension, or no tension at all, in response to certain stimuli, her behaviour changed. She became more disobedient! And although this sounds like a paradox, it is proof that she is on the right track. Patricia expressed this by pulling Esther along to places she did not want or was allowed to go (a section of the owner's private property that was behind the stable). So, disobedience: hurray!

Do not get mad at your horse for this disobedience; instead, come up with a way to deal with it that does not encroach on the established goals or methods. Esther and Patricia did my haystack exercise. This exercise contains a number of goals at the same time: it works on the relationship between the horse and his handler, builds the reciprocity between them, and introduces two signals to invite a horse to come along and to make him stop. You can find the haystack exercise in the Appendix. We did this exercise twice, each time on a different day. By then, the communication between Esther and Patricia had become a bit more in sync and their reciprocity and listening to one another had normalised. Because Patricia became so relaxed, there was no need for further repetition of the exercise. It also helped that Esther kept using the sound and gesture for 'go' and 'stop' during other activities in daily life.

What Esther thought of this approach:

I really like learning to understand Patricia better. Because Rachaël showed me which body features and signals Patricia was using in different places on the grounds, I was much better able to interpret her behaviour and level of tension. I feel like I gained tools that I can use to help her. This gives me a feeling of understanding for Patricia, but also of control over the situation. I really like it that Patricia did not experience pain in any way, and that it was not necessary for me to cause more tension in her. I would really have hated that. I do sometimes find it difficult to build an enriched environment. I am afraid of what my stablemates might say.

Note: When we got to this point in the behaviour adjustment plan, Esther got a new job, which meant Patricia also moved to a new stable. There, the layout is such that she can see all the horses from her stall and when she is being saddled or ridden in the arena. If Esther wants to take rides in the country, she can do this together with other riders. If Patricia had stayed at the same stable, then we would have stayed the course and expanded it, making the green zones even bigger and creating a situation in which Esther could have taken Patricia exploring, on rides, and on walks outside stable grounds as well.

4.3 THE ENRICHED ENVIRONMENT AS A MEANS TO BRING THE HORSE INTO BALANCE MENTALLY AND PHYSICALLY

Exploring and tracking are the more natural things a horse can do. However, if you have a horse who still needs to learn a lot about the human world, or who needs to discover it again in order to attach positive emotions to it, you will start by charting your horse's living situation. This has to meet the criteria of the freedoms described in section 1.2. You also want to design your horse's life in such a way that he can live inside his own comfort zone. In addition, you want to use exploration and scentwork (in this case, a treat search) to bring the horse into balance enough that he is mentally ready to start practicing the more human demands, such as lifting up his foot on command, walking on a rope in a certain way, having a halter put on him, etc.

4.4 THE ENRICHED ENVIRONMENT AS STAND-ALONE EXERCISES

Scentwork enriched environments can be a useful tool when you want to occupy or distract your horse.

Indy had just been moved to our stable. She was three years old and found a lot of stimuli intimidating. She was at pasture with Vos when I chose to lay the large black cloth down in the pasture with some feed pellets on it. She found this pretty scary, and even though Vos gave the right example by eating the pellets, it took her another three sessions to pluck up the courage to step onto the cloth herself to find the food. After that, however, she loved the cloth; she can now spend as long as 20 minutes trampling around on it, which is exactly what we wanted. Indy used to have trouble being separated from Vos. When I went to do something alone with Vos and placed her in the paddock, closest to us beside the arena, she did not handle it well. She kept pacing and looking at us. The black cloth was a good go-between. I placed it in the paddock with the feed pellets on it, readied Vos close to her, which kept her in her comfort zone, and then brought her to the paddock at the same time that I led Vos into the arena. In the 20 minutes that Vos and I were in the arena, she was busy searching on the cloth. This turned out to be a very good distraction, and the effect of the searching was also positive: it calmed her down. That way, even when we were not done yet, she stood looking at us calmly, rather than anxiously walking around.

HOW SUCCESSFUL CAN IT BE?

This method is very successful if you make substantial room for it in your life and work with horses and train with it at least twice a week. This means you need to have a different perspective on training and start to view exploration and tracking as equally valuable and indispensable training elements. Susan Kjærgård, who has embraced this technique in her life and work with horses, has proven that this is possible at a high level of training. Susan has showcased horses for 14 years on Denmark's national team in equestrian jumping and competed on countless horses around the world. Her highest ranking on the world rankings was number 76, and she is proud that all her Grand Prix (GP) rankings have been achieved on horses she has trained herself from scratch. Susan is the founder and driving force behind Blue Berry Hill, an education programme in Denmark, to become a riding instructor with a very high standard of teaching.

Susan: 'I have been out of the competition world for almost 10 years. Years I spent building Blueberryhill, a school for riding instructors, but also years I spent figuring out how to balance the life and demands of a competition horse with my desire to give the horse a life that is worthy and meaning full to him, outside of the results he produces for human gain. The answers to that process took shape and led me again to the competition ring. Rachaël's methods play an important part in this. My training has been given a whole new dimension, and the effect of the training has been mind blowing with each of the highly sensitive competition horses (of various disciplines) I am working with.

One of these is a sensitive and highly talented mare, Syvhøjegårds Clara. She lost the motivation to jump in bigger classes with different riders. I am working to bring her from a low level back into a high level with confidence, initiative, and high motivation. I have been allowed a nice timeframe and space for "alternative training sessions", i.e. mental stimulation and enriched environments, as part of the regular training (**Fig. 4.23**).

Fig. 4.23

With her, my initial goal was to motivate her to be super interested in both me, our joint relationship, and work in general, so I combined days of bodywork and training in hand and under saddle with lots of shorter sessions of mental stimulation.

My first step was a combination of an enriched environment and treat search. I had a few things, such as a towel, a bucket, and a dog basket put in the arena and filled it with food. At first, she was scared and hesitant, but she is highly motivated by food, so when she found the food, she calmed quickly.

She had a fearful association with the blue plastic liverpool (water hurdle) so we spent a lot of time investigating it, eating out of it, and stepping onto it and over it. From the saddle, I would walk up and throw a few treats into it, which she would eat. Now she jumps it easily and confidently.

Now, I use more and more advanced treat searches, like a small kids' pool filled with empty plastic bottles to make noise, which I then have her eat her breakfast out of. (**Fig. 4.24**). She can now open boxes and get treats out of bags, and she is curious and ready to investigate every new stimulus. On a trail ride, she saw a jacket on the side of a road, stared at it, and then walked right up to investigate it.

Fig. 4.24

Rachaël's exercises described in this book have really helped to bring out the curiosity and positivity in this horse. She investigates a lot with her tongue and teeth, enjoys licking the jumps in the arena, and has her nose everywhere. We now have a signal for when to go explore and for when to "be on the job", and I find that her attitude towards work is very positive and confident.

To riders with a more traditional mindset, she might seem "spoiled" and annoying because she wants to be part of everything and sticks her nose up in people's faces. Some interpret this as disrespecting their personal space. To me, it is part of her personality, and because I give her the chance to express that, she has become much better balanced mentally and easier to train and jump with.

I feel like the mental stimulation work makes up for some of the acute stressors she is dealing with in life as a competition horse, and I adapt the "Rachaël sessions" to fit into the weekly schedule.

Especially the change in my attitude from having all behaviour happen on cue to making the horse become more independent and take initiative is having a profound effect on her resilience.

This is what I do:

- She has physical rest days during which we do treat searches and/or enriched environments.
- We go for walks during which she decides the route (hand walking is good for building topline).
- I have her in extra-long and loose ropes for grooming, so she can have control and move to show me the good spots.
- She gets to steal hay when passing the wheelbarrow.
- We have breaks when riding during which she's in charge of what we do.
- We start and finish riding sessions by hand, and spend a little time walking around investigating while warming up or cooling down (which needs to be done anyway).
- Before loading for a show, I do a short treat search session to encourage initiative and boost dopamine (**Fig. 4.25**).

I have noticed the following benefits:

- She comes to me willingly in both the grass pasture and the box.
- She is positive and friendly to all the people walking by her.
- She is braver and able to handle unexpected stimuli.
- She can concentrate for longer, even when it's windy.
- She enjoys grooming.
- She jumps the water jump without hesitation.
- She is processing feed well and maintaining weight on less grain (still lots of hay).
- She is more patient, i.e. standing still for tacking up at shows.
- Less aggression towards other horses.
- Overall performance inhancement.

I jump her in a plain snaffle, with a loose noseband and no spurs, and she is very willing. For me, it's so important that a showjumping horse has initiative and problem solving skills. You're sure to need that at some point when jumping gets challenging.

I have always believed that riding is a partnership, not a dictatorship. Rachaëls approach, in which, to a significant extent, you allow the horse to also bring his ideas to the table, has prompted me to adjust my training, and I'm loving the results

Fig. 4.25 Susan set up a treat search (more on that in Chapter 18) on the liverpool to get Clara into a positive state before the horse show begins.

of improved show performance AND greatly improved enjoyment for the horse (and myself).

I have noticed that people have started to gravitate towards Clara: everybody wants to pet her and talk with her, and she is developing a small "fanbase". I think they are instinctively feeling her personality and contentment and are attracted to that' (**Fig. 4.26**).

Fig. 4.26 Susan and Clara in training.

Part 2: Scent tracking for horses

5 *Introducing scent tracking for horses*

5.1 USING SCENT IS VITAL FOR A HORSE

Sight, hearing, taste, touch, smell: The use of the senses is crucial to a horse. It is a matter of survival and wellbeing. This starts in the very first moments after birth. The foal who can use his senses to find his mother's teat survives. The horse who can use his senses to stay with the herd or find them again after he has lost them regains social security, connection, and safety. Because he is back in the herd, he can express the essential natural behaviours that are so important to his health and wellbeing, like lying down for undisturbed rapid-eye movement (REM) sleep while another horse stands guard.

Use of the senses, especially the sense of smell, spurs horses to immediate movement and action. A good scent assessment allows them to find suitable sexual mates, recognize which plants are edible and which are poisonous, or know what water they can drink. They can also recognise places or detect stimuli that might be a danger to them, such as the scent of an unknown animal close by or smoke in the air. The nose plays a crucial role in all of this.

When we are talking about the senses, the nose and smelling are not the first things that come to mind to us humans. We are a little surprised when we see a video on social media in which a dog greets his owner, whom he has not seen in a long time, but he does not react very enthusiastically at first when he sees and touches him. It is not until the dog recognises his owner by her scent and then gets intensely happy that we say, 'right, that had to do with scent: he is only now smelling who it is'. It is not like we realized this beforehand. Similarly, the idea that a horse who is startled by seemingly nothing might do this because he was spooked by a scent or part of a scent he suddenly recognises as negative or dangerous is a bit foreign to us.

It's not that smell is not important for humans too, as we subconsciously, and sometimes consciously, use it a lot (see section 6.6). But I do believe we use it to a much lesser extent than horses and dogs do. Scents have a different meaning to every species. To a predator or a scavenger, the scent of rotting meat is alluring, tempting them to approach it to find a meal. Horses, on the other hand, will avoid this scent, because there is a chance they will encounter a predator at its source. If we smell them at all, I think people have a far less negative association with countless scents than horses do. In that respect, we have some catching up to do in order to understand all the aspects of the horse's nose and to grasp the role it plays in his life.

6 Anatomy and physiology

If you go by the information in books about horses with regard to the power of their sense of smell in comparison with their senses of touch and sight, their sense of smell does not come off well. I am a fan of practical facts and had hoped to find information like the distance at which a horse can smell a certain scent, how old a scent could be for a horse to still be able to smell it, and how much a compound can be diluted before a horse will no longer be able to smell it. And yes, I realise that these are questions that pertain directly to my scentwork with horses, but still: is there nothing practical available on the horse's sense of smell at all? It turns out that there is. In section 6.6, I name a few interesting studies that deal with the connection between the horse's sense of smell and his associated behaviours. When you start to do tracking with horses, I think it is important to have a theoretical framework for this as well. However, if you want to dive straight into the hands-on stuff, you can skip to Chapter 10.

6.1 THE INITIAL SMELLING SYSTEM: FROM SCENT TO ELECTRICAL SIGNAL IN THE BRAIN

The normal ambient air we humans and animals breathe consists of nitrogen (over 78%), oxygen (over 20%), and other gasses, such as argon, carbon dioxide, helium, and neon. In addition, water vapour is also a regular component of air. In a regular climate of roughly 20° Celsius, this makes up about 1–2%. In a very cold environment, this can drop to 0.02%, and in a tropical environment, it can rise to 4.4–6%.[1, 2] The gasses I just listed are all fairly constant. The following things vary in their concentration and manifestation: pollen, bacteria, sodium chloride, and spores, among which are scent traces.[3]

The horse does not breathe through his mouth but through his nose, so he draws in these scent traces and scent particles when he inhales air through his nostrils. Not every miniscule scent particle floating through the air is the same. Scent traces can vary in weight and concentration. Their volatility from liquid into gas is also a distinguishing feature.

There are volatile and nonvolatile scent compounds. The volatile compounds vaporise more quickly into a gaseous form than nonvolatile ones. For instance, in volatile compounds, this happens at room temperature, requiring no additional heating. Compare it to a bottle of acetone used to remove nail polish. This is kept in a bottle in liquid form. When the bottle is opened, some of the compound evaporates. We can smell this acetone very clearly when this happens. A compound that is nonvolatile requires a much higher temperature to vaporise and become gas. It can do so, but this requires more effort, for instance, by boiling it and adding pressure. The higher the temperature required to make a compound volatile (and turn it into gas), the heavier the nonvolatile compound is. Nonvolatile compounds are, for instance, sugar, salt, protein, and starch.[4]

Another example of this can be seen when a spoonful of sugar is added to a glass of water and stirred; the sugar will dissolve in the water, but it does not evaporate. That is why you cannot smell the sugar (sugar is a nonvolatile compound: it does not evaporate at room temperature). If, however, you mix some lemon juice into a glass of water, you will smell it. This compound is volatile: It evaporates at room temperature and floats through the air in its gaseous form.

What journey does a scent particle make?

STEP 1

The volatile scent compounds, which have a molecular mass of less than 300 Da (daltons),[5] represented by the red, purple, and green dots in the illustration, are sucked into the nasal cavities through the nostrils along with the air. When this happens, the nostrils can flare, causing a larger quantity of air to be inhaled. The air then continues upward through the nasal cavity. The nasal cavity is lined with epithelium, which makes the air that is breathed in more moist and warmer. The epithelium, aided by a layer of mucus, also functions as an immune barrier that protects the horse from harmful compounds, such as pollutants, allergens, and micro-organisms[6] (**Fig. 6.1**).

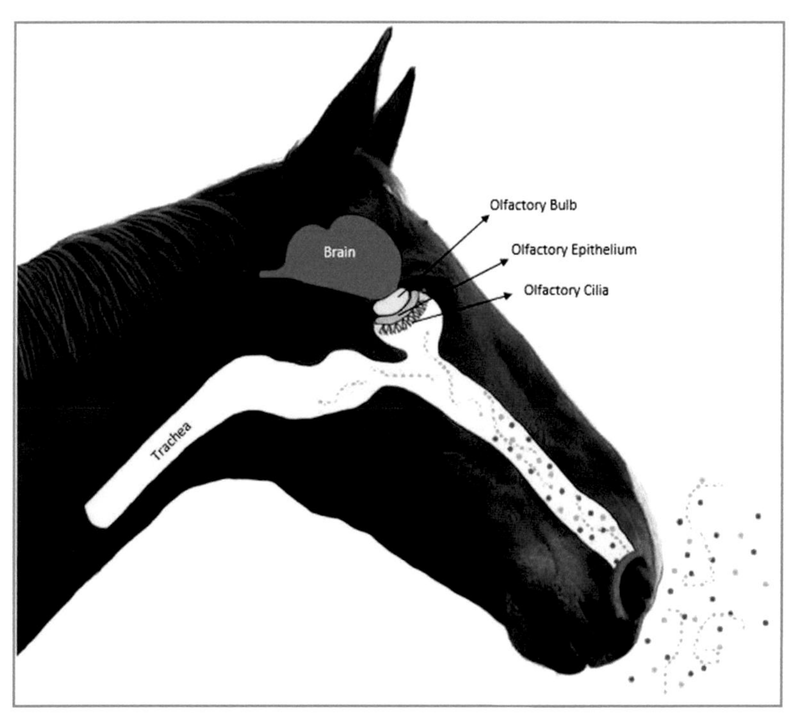

Fig. 6.1

STEP 2

The scent compounds stick to cilia. The cilia are hairs that are on the dendrite of the odorant receptor neuron. These dendrites with cilia dangle in the upper nasal cavity in a concentrated layer of mucus. Here, an initial sorting takes place in which certain scent molecules stick to specific cilia and their dendrite. If the chemical scent molecule reaches the dendrite through the cilia, then it is converted into an electric signal[6, 7] (**Fig. 6.2**).

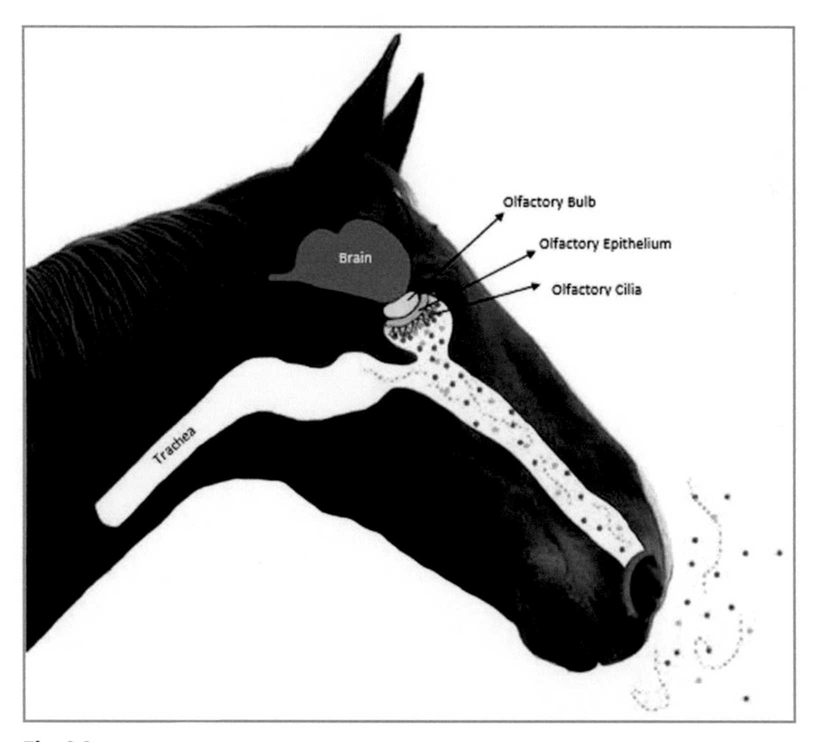

Fig. 6.2

STEP 3

The scent, as an electrical signal, continues its way up toward the olfactory receptor nucleus itself (**Figs. 6.3–6.4**).

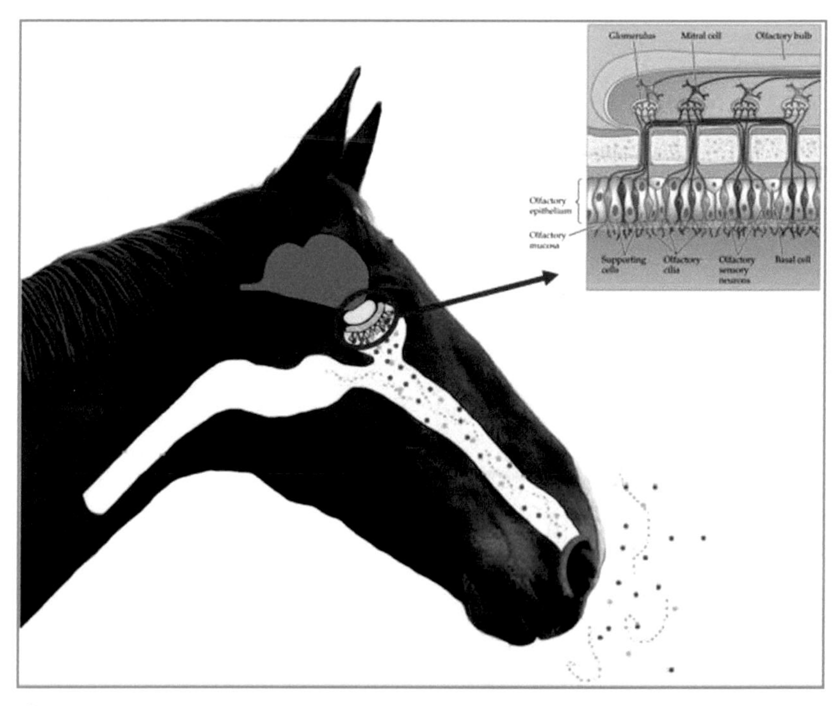

Fig. 6.3

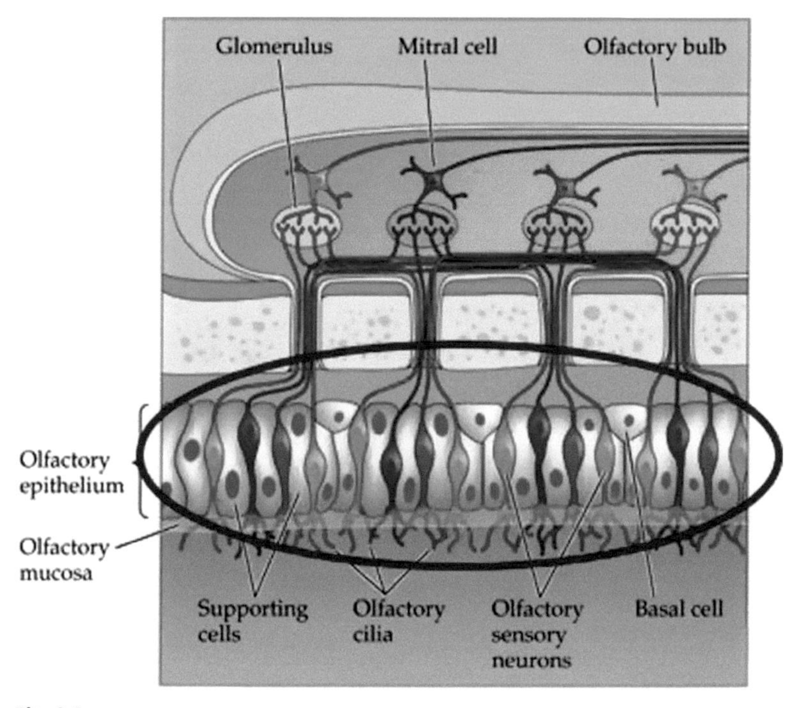
Glomerulus
Mitral cell
Olfactory bulb
Olfactory epithelium
Olfactory mucosa
Supporting cells
Olfactory cilia
Olfactory sensory neurons
Basal cell

Fig. 6.4

The olfactory reception neurons are bipolar neurons. The cell body has an axon on one side and a single dendrite on the other. Olfactory reception neurons are replaced throughout their lives.[6]

STEP 4

As electrical signals, the different scents are sorted by the olfactory receptor neurons and sent to their corresponding glomeruli. The olfactory receptor neurons do this by bundling a number of their matching axons, creating olfactory nerves that send the bundled electrical signals to a corresponding glomerulus through the ethmoid bone. Passing through the ethmoid bone, the scent goes from the nasal cavity (outside) to the brain (inside) (**Fig. 6.5**).

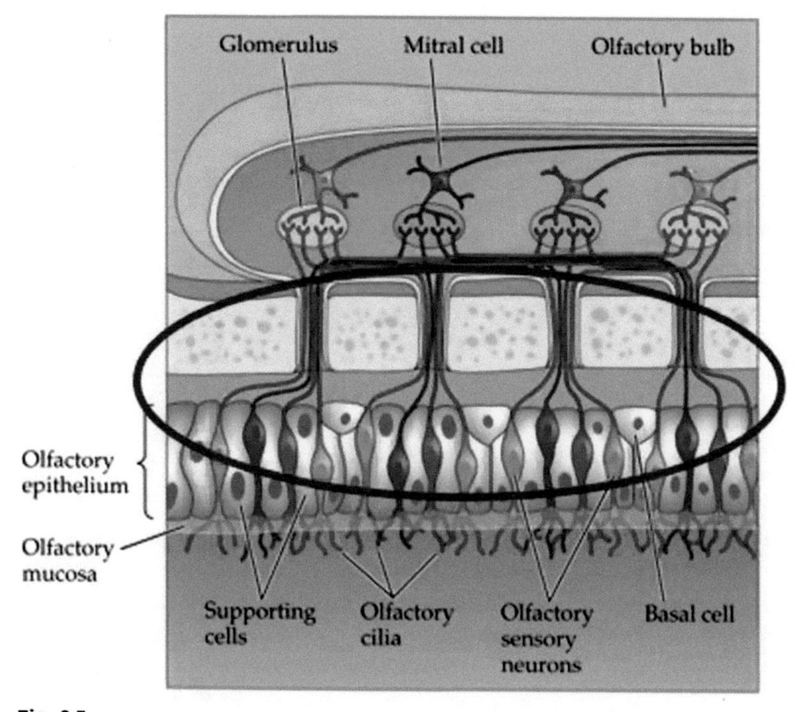

Fig. 6.5

There are thousands of glomeruli in the olfactory bulb that are structured for scent. Some respond to fatty acid, some to alkyl amine odorants, etc. Every glomerulus contains axon terminals of olfactory receptor neurons and dendrites of mitral and tufted cells (not shown in the illustration, **Figs. 6.4–6.6**).

STEP 5

The axons of the mitral and tufted cells form bundles, together constituting the olfactory tract. Scent information is sent to the brain along this tract.[5]

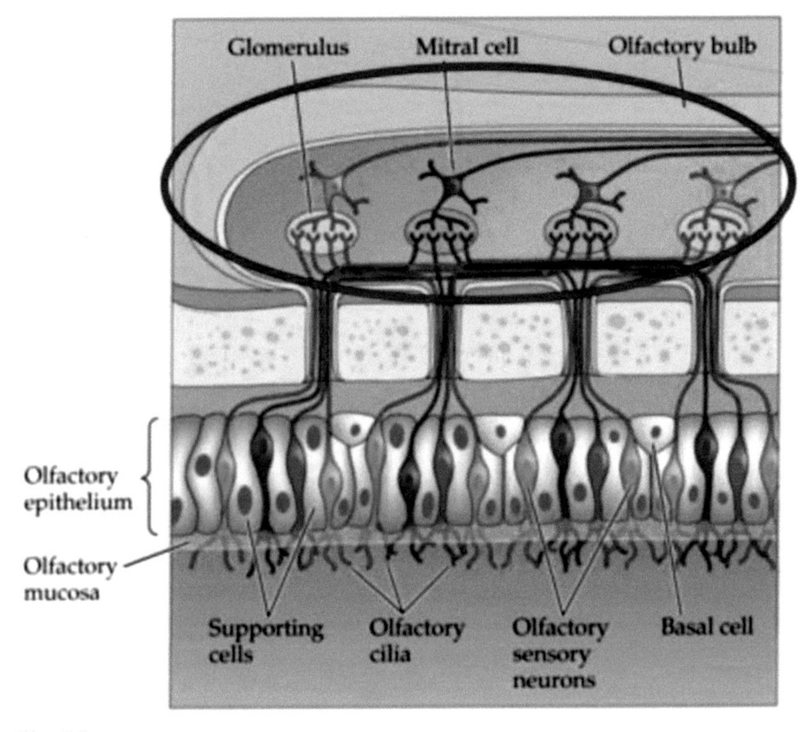

Fig. 6.6

The described olfactory route happens on both sides of the horse's head, so the horse has two identical olfactory systems, one on each side of his head. The scent particles that enter through the left nostril are also processed by the olfactory epithelium and olfactory bulb on the left side of the head, and they are sent to the brain along the left olfactory tract as well. The same thing happens on the right side of the head. This makes the olfactory system the only brain system that does not do a crossover.

STEP 6

The electrical signals are spread to the olfactory cortex along the olfactory tracts. The olfactory cortex contains the areas of the brain that receive direct afferent input from the olfactory bulb.[5, 8] This includes the olfactory tubercle, the pyriform cortex, the amygdala and the entorhinal cortex.[9, 6, 5] Next, the signals continue and are distributed to the orbitofrontal cortex, the thalamus, the hypothalamus, and the hippocampal formation.[5] There is also communication between the different areas.

The first four areas that are reached are the olfactory tubercle, the pyriform cortex, the amygdala, and the entorhinal cortex (**Fig. 6.7**). The olfactory tubercle plays an essential role in focusing on something and wanting something. It is related to desire and reward and wanting to approach something or not.[5] As such, the olfactory tubercle causes behaviours to follow in response to a scent.

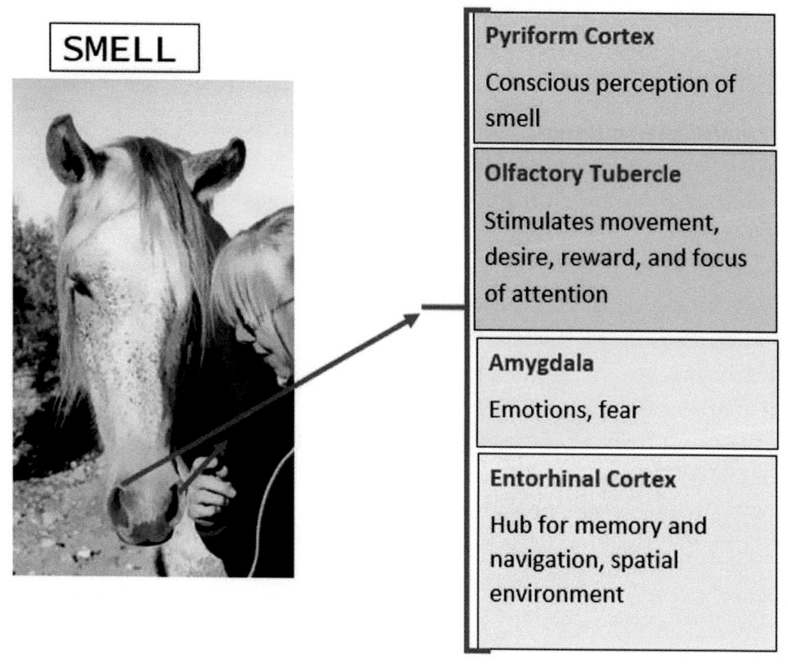

Fig. 6.7 Images and words are based on the work of Daler Purves, George Augustine, et al., as described in their book ***Neuroscience***, Oxford University Press, 2012.

In the pyriform cortex, awareness of the scent takes place. Here, the scent of the apple or the freshly baked bread is identified and consciously experienced. Together with the other three areas, it also causes emotion and association to be linked to a scent.[6]

THE INGENUITY OF THE OLFACTORY SYSTEM

Let us make the whole thing even more complicated. I have just described the journey a scent compound takes. I hereby imply that one scent compound is also the whole of one scent, that, for instance, the scent of apple consists of a single, comprehensive scent compound of one apple. Nothing could be further from the truth. The recognition of a single scent takes hard work, because the olfactory system has to process all the 300 scent molecules[10] that together form the scent of apple. (To give you another example, 350 scent molecules have been counted that together constitute the scent of real vanilla beans.[4]) When you inhale all the scent molecules of an apple through your nose, all 300 scent molecules follow the described olfactory route. In the pyriform cortex, these 300 scent molecules are identified and consciously recognised as one scent of one apple. You might think it is very inefficient to have to process this many scent particles (it seems easier to just register one scent molecule for a single vanilla bean), but from an evolutionary point of view, this system is particularly functional. With it, the olfactory system can recognise and process thousands of scent combinations.

At the same time, a signal is sent to the amygdala, the part of the brain that plays the crucial role in calling up emotions such as fear, anxiety, and aggression.[11]

We should not forget that the entorhinal cortex also receives olfactory information. The entorhinal cortex plays an important part in creating a cognitive map. This is a spatial map of the surroundings that includes personal experiences as well as information from the other senses. This allows you to navigate within an area. For instance, the horse knows where he is in his living environment, and where he needs to go to, for instance, to find water.[12] In my experience, the horse's cognitive map is extremely well developed. If a horse is unable to find a scent bag, you will see him return to the many places where he has found a reward before in a highly methodical fashion. He can even return to places where he found a scent bag as long ago as 2 weeks (making me wonder whether horses can still smell scent traces in these old tracking locations).

The next four areas that are also reached are the thalamus, the hypothalamus, the hippocampal formation, and the orbitofrontal cortex (**Fig 6.8**). The thalamus can be

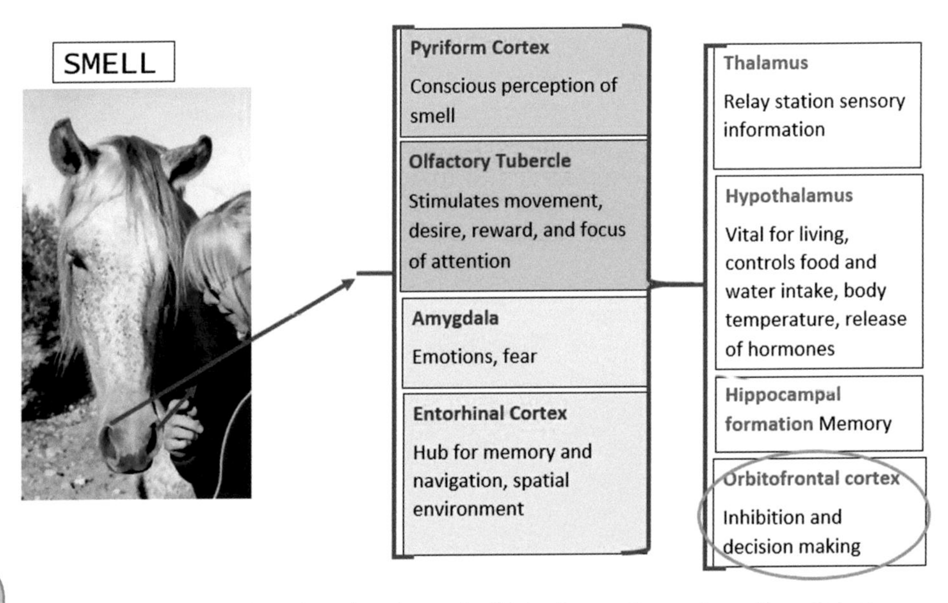

Fig. 6.8 Images and words are based on the work of Daler Purves, George Augustine, et al., as described in their book ***Neuroscience***, Oxford University Press, 2012.

viewed as the control centre for sensory information[6] in which the thalamus sends the stimuli from all the senses—such as the perception of temperature, touch, pain, sounds, or images the horse sees—to the different parts of the cerebral cortex. (The cerebral cortex, in turn, causes the signals it receives from the thalamus to be converted into concrete instructions to the body and the formation of mental images and thoughts.[13]) The thalamus also plays a role in the perception of the posture your body is in.[14]

The hypothalamus is a control centre that is indispensable to life. It regulates a large number of unconscious body functions. For instance, the hypothalamus regulates how hungry and thirsty you are, your body temperature, your day and night rhythm, and the release of hormones. The hypothalamus gets its information through constant feedback from the spinal cord about the different organs in your body. The hypothalamus also measures whether you have enough of a particular hormone in your blood. It directs the pituitary gland, which in turn releases more or fewer hormones according to what the body needs.[15]

The hippocampal formation is a subject of discussion among scientists when it comes to the areas of the brain that constitute it. They agree about one thing though: it includes the hippocampus.[16] This is an important component, because it is viewed as the place where memory is stored. It also plays an important part in our thoughts and actions. Thanks to the storied memories, a connection can be made between the experiences and thoughts we had in the past and the ones we are having now. So there is interaction between the past and the present, and the hippocampus is strongly involved in this.[17] This also includes humans' ability to think about themselves as individuals and all the little reminders that go with this. To what extent horses are able to do this I think is still a matter of debate. The same is true of the next area I will discuss: the orbitofrontal cortex.

The orbitofrontal cortex plays an important part in humans' ability to make decisions, display inhibitory behaviour, and recognise interpersonal signals and their emotional significance.[18] The latter is important in order to function well socially. The circle drawn around the orbitofrontal cortex in the illustration (**Fig. 6.8**) implies that this area of the brain is clearly visible in humans/primates. In other animals, it is of a different size or it does not exist. This has led to a heated discussed about whether animals have metathoughts, make decisions, are able to suppress impulses, and can read individual emotional signals. In this context, I want to refer to Frans de Waal's book *Are We Smart Enough to Know How Smart Animals Are?* Besides many interesting thoughts and musings, this book is full of examples of animal behaviour we used to think were reserved to the more 'intelligent' humans, like using tools, predicting what will happen in the future and adapting your plans accordingly, and the ability to delay gratification for a larger reward down the line. The brains of mammals are mostly alike: I think that the lack of an orbitofrontal cortex does not mean that an animal is not be able to show the behaviours above. A crow does not have a neocortex, but crows are great problem solvers. Dolphins have a small size orbitofrontal lobe, and they are highly intelligent, socially rounded beings. In my view, this is also true of horses. And based on my own experience,

I can assert that, when they know they can, they do make decisions. Better yet, the more opportunities and encouragements they have, the more they are able to recall past decisions and the more confident they are in trying out new decisions. They are also very capable of reading the interpersonal signals of others.

6.2 THE UNIQUE CHARACTERISTICS OF 'SMELLING'

A question that came up several times during my childhood when I would play with my cousins was 'if you could choose, would you rather be deaf or blind?' I now realise we had not even incorporated smell into the equation. This when the sense of smell does have some unique features compared to sight and hearing. I will name two.

1. The sense of smell, and with it tracking, searching, and discovering, is distinguished from hearing and sight in that the olfactory system is densely wired with dopaminergic neurons and driven by the neurotransmitter dopamine.[18, 5, 18] For instance, you can look forward to hearing a particular song, and that will release dopamine, but the amount of dopamine realised in targeted searching is many times higher, and cannot be compared to the small amount released when you want to hear a song. Searching and tracking is fed and stimulated by dopamine.[18]

2. Compared to sight and sound, scent travels its own unique route in the brain. When you hear or see, the impulses first go to the control centre, the thalamus, which then distributes the impulses further. When it comes to smell, as we have seen in **Fig. 6.8**, the first impulses go to the pyriform cortex, olfactory tubercle, amygdala, and entorhinal cortex, and only then do they go to the thalamus.[6] To compare: in a highly simplified representation of sight and hearing, it looks like this: **(Fig. 6.9)**

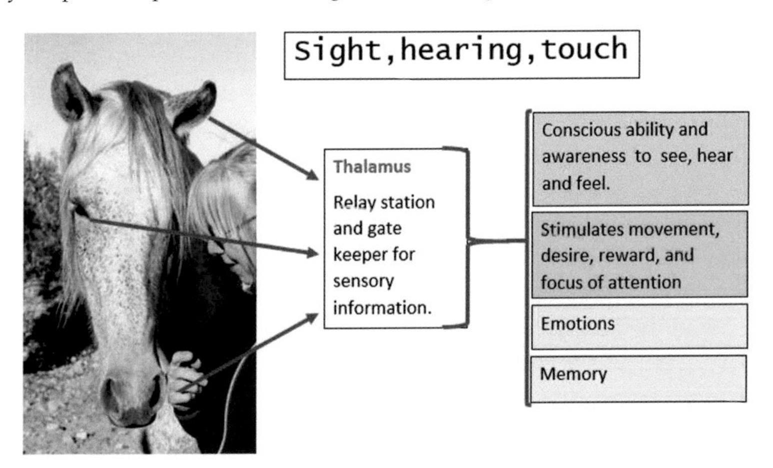

Fig. 6.9 Images and words are based on the work of Daler Purves, George Augustine, et al., as described in their book ***Neuroscience***, Oxford University Press, 2012.

Perhaps the differences described above can be explained by evolution. Maybe 400 million years ago, when fish-like animals first crawled onto land, olfaction was the only sense that helped them to gather information from a distance. In addition, in this first mammalian brain, olfaction was the primary source of information for the limbic system, which is involved in emotions, motivations, and memory.[19]

In conclusion, the horse's sense of smell plays an important part in his physical system. It is designed to strongly motivate and drive a horse to take actions. The olfactory sense catches scents, processes them, and remembers them. Scents are coupled with emotions, general memories, and visual spatial images of the environment.

6.3 ARE THERE ALSO HORSES WHO CANNOT SMELL?

In people, the sense of smell declines as they get older. Until we are 40, we are capable of distinguishing 80% of common scents. After that, the ability declines, with this downward trajectory continuing as we get older.[6] I have not found any studies in which it was specifically tested whether the same is true of horses. Personally, I have yet to come across a horse who could not track because of his age, though it must be said that the oldest horse I have tracked with was 30 years old. It is possible to find an older horse, of course! Perhaps an even older horse would show a diminished sense of smell. At the moment, this is just speculation. Of course, the sense of smell is a crucial sense that helps the horse to survive. I can imagine, therefore, that the olfactory system remains intact for a long time, or at least intact enough to keep the horse alive.

In people, chemotherapy, traumatic head injury, and Alzheimer's disease can very negatively impact the sense of smell.[6] Another common cause of deterioration of the olfactory sense in people is an infection in the upper and/or lower respiratory tract. The increased production of mucus makes it more difficult for oxygen and scent particles to be absorbed and processed. It also often causes shortness of breath. In horses, a similar mechanism can occur. There are horses who have infections of the upper respiratory tract (nasal passages, throat, and airway) and the lower respiratory tract (bronchi, bronchioles, and alveoli in the lungs). They can also contract bacterial or viral infection, or have allergies like bronchitis, asthma, lung emphysema, or COPD. I have not done studies on the effects, positive or negative, of tracking for these horses. The horses I have tracked with who had any of these ailments can be counted on two hands, which is not enough to base any kind of conclusion on.

One of the best things about tracking is that you cannot force your horse to do it. Whereas you can pressure your horse to do what you want in dressage, jumping, or any other discipline (possibly by use of spurs, a whip, or whirling ropes), you cannot do so in tracking. You cannot push your horse's head down and command him to 'smell'. If you put pressure on the horse in this case, he would sooner run than track. As such, your horse will demonstrate of his own accord whether or not he can handle this. A note on this: an incidental cold is, of course, very different than a chronic ailment. If your horse has this, I would consult with your vet or a specialist first. Because it is possible that your horse inhales air more deeply during tracking than he normally would. His lungs and midriff expand when this happens. For instance if your horse has pulmonary emphysema, he already has trouble letting the air he has breathed in flow back out (because of the increased build-up of mucus). I can imagine this will only get worse during tracking. With a horse like this, I would choose to do more enriched environments, rather than tracking, as described in Part 1.

The study 'Reduction of the Olfactory Cognitive Ability in Horses During Preslaughter: Stress-related Hormones Evaluation' by Micera, et al., looked at the relationship between scent and tension and how this is influenced. The researchers studied horses who were being brought to the slaughterhouse to be slaughtered. They assumed that horses feel tension when they arrive there because they perceive several odour signals of danger. When, however, the horses from the experiment group had a mentholated ointment smeared into their nostrils 45 minutes before being slaughtered, 'their adrenergic response to the slaughterhouse was reduced, implying that this method may reduce horses' stress response'. It demonstrates that the scent of the mentholated ointment itself did not raise the horse's stress levels in that setting, but that it was strong enough to block out the other, 'dangerous' scents. It would be interesting to investigate the extent to which nose gels containing pheromones against stress or vaporisers/inhalers that administer medicines through the nose can influence tracking, as it seems probable that they would have an impact. To what extent and for how long is an open question.

6.4 WHAT DOES THE SECOND SMELLING SYSTEM DO? THE VNO/JACOBSEN ORGAN

The previously explained olfactory system is also referred to as the first olfactory system. The second olfactory system is the one that includes the Jacobsen organ, also called the *vomeronasal organ*. This second olfactory system is specialised in seeking out and processing nonvolatile compounds of organic origin, such as hormones and pheromones. Pheromones are scent particles that provide information about gender, age, health, and reproductive status.[11] This should not be confused with signature scents; these are scents that can give one horse information about individual characteristics of another specific horse. For instance, if you let people smell the t-shirts that belong to other people, those doing the smelling will have a preference for a specific t-shirt. They base this personal choice on the signature odours of the wearer of that shirt. This can be the scent of sweat, saliva, deodorant, etc. However, signature scents do not provoke within-sex or between-sex behavioural or endocrine changes, while pheromones do.

Fig. 6.10

Hormones and pheromones, as nonvolatile compounds, have a molecular mass of over 300 Da (daltons); they do not vaporise easily. That is why horses inhale these compounds with a lot of force (**Fig. 6.10**). Because of the force of the inhalation and the pressure it causes, the compounds dissolve and are able to be processed by the olfactory system and the brain. Although it is interesting, this second olfactory system does not play a role in my tracking method. I will leave it aside for the rest of the book.

6.5 WHAT DOES IT SOUND LIKE?

During tracking, a horse can make a very recognisable sound: a soft snorting. This is an especially effective behaviour, because the snorting empties the horse's nasal passages, enabling him to inhale more air and smell better. Horses can do this while they are tracking. Some horses who know they are going to be tracking empty their nasal passages before they get started, as if they are preparing their instrument for the task ahead.

Another characteristic sound you might hear when your horse is tracking is the audible, repetitive deeper inhalation of air into the nostrils. For this, the nostrils repeatedly expand and contract. You can also sometimes hear these deep inhalations cause vibrations inside the nasal passages. This sounds like a continuous low kind of snore with vibrations in it (for an example, you can go on YouTube and find a clip of mine called *A Snort and a Scar*). In some horses, I can hear that they have this constant, low, vibrating snore when they are working on an especially difficult task, or when they are investigating a scent for a longer time with persistent high intensity. This is not true of all horses though. However, I do hear all of them snorting to clear their nasal passages once they have tracked a few times and are motivated to do this.

6.6 WHAT DOES THIS MEAN FOR EVERYDAY LIFE?

Is the colour green that I see the same colour you see? Is what I smell the same as what you smell? How do you get a sense of these abilities, like the degree to which your horse is able to smell?

Because most people have a feeling and an idea of what they can smell and how a dog uses his sense of smell, I decided to see if I could make a comparison between humans, dogs, and horses. I went in search of a common component that would provide information on all three in order to place them alongside one another. For instance, I went in search of the surface area of the nasal mucosa and the olfactory epithelium, the estimated number of glomeruli and axons that travel from the olfactory neurons to the glomeruli, and the size of the olfactory bulb. It turned out to be a very challenging quest. A lot of information is available about humans and dogs, but not nearly as much about horses. And if I found encouraging numbers about, for instance, the numbers of olfactory cells horses have, I could not find the research the claim was based on, making me unsure as to how reliable the information was. I did find studies by Ramesh Padodara,[20] Laurie Issel-Tarver, Jasper Rine,[21] and Yoshihito Niimura, et al.[22] They did research into the olfactory abilities of various animals, including the horse. They compared the horse to dogs and humans specifically in terms of the number of intact, functional olfactory receptor genes. Below, I share Nimimura and Padodara's numbers. I also use the numbers for humans and dogs I found in the book *Neuroscience*.[6] The numbers that reflect the size of the olfactory bulb of people and horses I found in the book *The Mind of the Horse* by Michel-Antoine Leblanc[12] (I am using the averages here), and those on dogs from the study *Comparative Morphometry of the Olfactory Bulb, Tract and Stria in the Human, Dog and Goat* by Boniface M. Kavoi and Hassanali Jameela (**Fig. 6.11**).

	Intact Olfactory Receptor Genes	Size of Olfactory Bulb
Human	396	11 mm long x 3.5 mm wide
Horse	1066	33 mm long x 23 mm wide
Dog	811-1100	19 mm long x 11 mm wide

Fig. 6.11

According to this particular comparison of functional olfactory receptor genes and the size of the olfactory bulb, the horse's sense of smell is comparable to the dog's. In the study by Issel-Tarver and Jasper Rine, the horse's olfactory ability is placed below that of dogs and wolves but above that of humans. I found this ranking more often in articles and books, but I was never able to trace their sources.

Some critical notes and observations I discovered are that it is uncertain whether the number of intact olfactory receptor genes and the size of the olfactory bulb are an indication of better olfactory ability. The size of the surface area of the olfactory epithelium might also play a role as well as the density of olfactory cells and their number. The way horses smell and how driven they are on certain trails I lay in the arena or the forest do make me place them on the same level with certain dogs I have done scent tracking with. The fact that these are two different species, each with their own behaviours and motivations, can create a distorted view in which the horse's sense of smell, because of his behaviour, might be viewed as inferior to that of a dog. It goes without saying that much more research is needed before questions like these can be answered.

WHAT CAPACITIES DO HUMANS HAVE?

It is difficult to form a picture on the basis of the abstract numbers in Table 6.11 and to understand what they mean for everyday use. That is why I will offer some examples of what the human sense of smell can do with its 396 intact olfactory receptor genes.

If a person knows the scent of ethyl mercaptan (this is the odorant that is added to propane as a warning agent), and you put her in front of two Olympic-size swimming pools, one of which contains three drops of ethyl mercaptan, she can point out which one it is.[5]

A study that is also often mentioned and that captures the imagination is that on 'the smell of fear'. In this study, people were asked to watch a movie. One group watch a funny movie, the other group a scary one. As they watched, their underarm odours were captured on gauze pads. These pads were kept in closed bottles. A week later, a group of men and women were asked to smell these bottles. There were bottles that did not contain any scent, bottles that contained the scent of happy people, and the bottles that held the scent of the people who had been scared. Those doing the smelling had to decide if they were smelling happy sweat or scared sweat. The women had a very high accuracy rate in detecting happy sweat in both men and women and detecting scared sweat in men. Men had a very high rate of identifying happy sweat in women and scared sweat in men.[23]

Another well-known fact I would like to include is the tendency for women's menstrual cycles to sync when they spend a lot of time together, for instance at work or in the home. How does this happen? Women unconsciously smell when another woman has entered her ovulatory or follicular period.[5] Pretty exceptional, right? I also found it interesting to read that we humans are able to strongly improve our sense of smell through training. If we ourselves get down on our knees and follow a scent trail of baby powder, after four days, we are apparently able the follow the scent as fast as we can crawl.[5] (Note: If you want to try this for yourself, use the scent of baby powder. This scent is very detectable to us. Have someone else lay a trail with this outside on the grass that you will follow with your nose. And, while you are at it, you can see if you get a dopamine rush.)

WHAT CAPACITIES DO HORSES HAVE?

In writing this paragraph, I realise how difficult (and frustrating) it is to barely be able to answer this question. Why? Because in daily life, you often just see the result of abilities you also have yourself. For instance, I will see that horses sniff my hands 3 hours after I have put on them a new lotion with Aloe vera. But when I smell my own hands and try hard, I can still smell the lotion as well. I can see that my horse can smell a person at a 100 metres distance. I can see his nostrils flaring, as well as the person whose scent he is detecting. But it also happens often that I see my horse catching a scent, but I have no idea where it is coming from. My own senses cannot detect what the horse is perceiving, making it impossible for me to chart these 'better' capabilities the horse possesses.

In addition, in real life, it is also difficult to rule out certain variables. When my horse spontaneously starts to follow another horse's trail in the forest, and I describe the route after returning to the stable, my stablemate might say, 'Hey, that is the exact route I took yesterday'. The question then is: was my horse following that 1-day-old trail, or had another horse ridden the same route that morning, creating a new and fresher trail? I cannot rule it out.

A horse can smell a drop of ketchup in a cement tub filled with water. Yes, this sounds amazing, but if we humans can smell which Olympic-sized swimming pool contains three drops of ethyl mercaptan, the drop of ketchup suddenly does not seem so impressive anymore.

That a horse's sense of smell is much better than a human's is clear to me. I deduce this, among other things, from the fact that a horse will stop tracking if the scent on the trail is too strong. These more concentrated scents are 'normal' scents to us. I personally believe that these strong scents literally fill up the horse's olfactory system so much that there is no more room for diluted scents (see Chapter 8). And it is these diluted, subtler scents, the ones we cannot smell, that they work so well with for a longer time.

I can imagine that if you want a practical comparison of the sense of smell between humans, horses, and dogs, you can create one by doing scent discrimination with your horse, in which you teach him to point to a particular scent. You could keep diluting this scent, enabling you to determine the point at which the horse can no longer smell it.

So, are there no studies about horses and their appetite for scents? Sure, there are. I found the book *The Mind of the Horse* by Michel-Antoine Leblanc a great help, and I also read a lot of studies on the internet. I have listed some of them in the bibliography. The purpose of these studies is not necessarily the horse's olfactory capacity, but more the identification of a link between the horse's sense of smell and his behaviour. For example, researchers looked into which dropping a horse was most interested in: his own, that of known herd mates, or that of horses who were unknown to him.[24] Researchers investigated the degree to which urine odour contains some information that horses can use to discriminate between conspecifics (members of the same species).[25] They also studied whether stallions defecate over mares' droppings to mask their scent;[26] how horses react to the scent of predators; and the behaviour and tension levels of a group of 2-year-old stallions being presented with new visual, auditory, or olfactory stimuli.[27]

REFERENCES

1. AirSain. (n.d.) *Air measurements, values in air.* https://www.airsain.nl/samenstelling-ideale-lucht
2. European Environment Agency. (2013) *With every breath.* https://www.eea.europa.eu/nl/ema-signalen/signalen-2013/artikelen/bij-elke-ademhaling
3. Lenntech. (n.d.) *Chemical composition of air.* https://www.lenntech.nl/lucht-samenstelling.htm
4. Maat L. (2007) *Very detailed.* https://www.nemokennislink.nl/publicaties/in-geuren-en-kleuren/
5. Kensaku M., et al. (2014) *The olfactory system, from odor molecules to motivational behaviours.* Springer Science+Business Media, Tokyo Japan.
6. Purves D., Augustine G., et al. (2012) *Neuroscience,* International Fifth Edition. Oxford University Press, New York.
7. Jenkins P., et al. (2009) *Olfactory cilia: Linking sensory cilia function and human disease.* https://www.ncbi.nlm.nih.gov/pmc/articles/PMC2682445/
8. Wilson D.A. (2009) *Olfactory cortex.* https://www.sciencedirect.com/topics/medicine-and-dentistry/olfactory-cortex
9. Merriam-Webster. (n.d.) Olfactory cortex. https://www.merriam-webster.com/medical/olfactory%20cortex
10. Espino-Diaz M., Sepulveda D., et al. *Biochemistry of apple aroma: A review.* https://www.ncbi.nlm.nih.gov/pmc/articles/PMC5253989/
11. Sapolsky R.M. (2017) *Behave: The biology of humans at our best and worst.* Penguin Random House.
12. Leblanc M.-A. (2013) *The mind of the horse: An introduction to equine cognition.* Editions Belin.
13. Wikipedia. (2019) *Brain cortex.* https://nl.wikipedia.org/wiki/Hersenschors
14. Hersenstichting (n.d.) *Anatomy of our brain.* https://www.hersenstichting.nl/dit-doen-wij/voorlichting/werking-van-de-hersenen/anatomie/
15. Nederlandse Hypofyse Stichting. (n.d.) *The hypothalamus.* https://www.hypofyse.nl/de-hypofyse/de-hypothalamus.html
16. Neuroscientifically Challenged. (2015) *2-minute neuroscience: The hippocampus.* https://www.youtube.com/watch?v=5EyaGR8GGhs

17. Hersenletsel-uitleg. (n.d.) *Hippocampus*. https://www.hersenletsel-uitleg.nl/gevolgen/gevolgen-per-hersendeel/hippocampus
18. Panksepp J. Biven L. (2012) *The archaeology of mind: Neuroevolutionary origins of human emotions*. W.W. Norton & Company.
19. Saslow C.A. (2002) *Understanding the perceptual world of horses*. https://www.sciencedirect.com/science/article/abs/pii/S0168159102000928
20. Padodara R. (2014) *The olfactory sense in different animals*. https://www.researchgate.net/publication/262932824_Olfactory_Sense_in_Different_Animals
21. Issel-Tarver L., Rine J. (1997) *The evolution of mammalian olfactory receptor genes*. https://www.researchgate.net/publication/14192454_The_Evolution_of_Mammalian_Olfactory_Receptor_Genes
22. Niimura Y., Matdui A., Touhare K. (2014) *Extreme expansion of the olfactory receptor gene repertoire in african elephants and evolutionary dynamics of orthologous gene groups in 13 placental mammals*. https://www.ncbi.nlm.nih.gov/pmc/articles/PMC4158756/
23. Chen D., Haviland-Jones J. (2001) *Human olfactory communication of emotion*. https://www.researchgate.net/publication/12175411_Human_Olfactory_Communication_of_Emotion
24. Kruger, K., Flauger, B. (2010) *Olfactory recognition of individual competitors by means of faeces in horse* (Equus caballus). https://www.researchgate.net/publication/49659137_Olfactory_recognition_of_individual_competitors_by_means_of_faeces_in_horse_Equus_caballus
25. Hothersall B., et al. (2010) *Discrimination between conspecific odour samples in the horse* (Equus caballus). https://www.sciencedirect.com/science/article/abs/pii/S0168159110001565
26. Kimura B. (2001) *Volatile substances in feces, urine and urine-marked feces of feral horses*. https://www.researchgate.net/publication/269637926_Volatile_substances_in_feces_urine_and_urine-marked_feces_of_feral_horses
27. Christensen K., et al. (2005) *Responses of horses to novel visual, olfactory and auditory stimuli*. http://epsilon.slu.se/avh/2006/jwc_paperI.pdf

7 *Biomechanics*

7.1 STUDY OBJECTIVES

A healthy brain in a healthy body is what we want for our horses. So, you naturally come to the question of what scent tracking does to the horse's body and what effect this might have. To study this, Nanne Boekholt, Julia Robertson, and I set up a study in which we watched footage of horses tracking. We posed the following research questions:

- What movements does the horse make when he is tracking?
- What happens to his body when he is tracking?
- What are the health benefits?
- What are the contraindications?

We tried to answer these questions in horses who were following a scent trail at a walk with a low to mid-low head/neck position. This could be a trail on the ground or tracking through air scenting. The horses who were following trails that were too easy for them, for instance, at a trot or with a mostly mid-low to mid-high head and neck position, we did not include in the study.

Here, we are mainly focused on what can be observed when you are looking at a horse who is tracking. What movements does he make? How does he hold his neck, back, hind leg? How mobile is he? Can he bring his nose down to the ground and maintain his balance while walking? In this chapter, we are primarily focused on the anatomy and biomechanics of the horse when he is tracking. In the future, we hope to add neurology and physiology to this. This is an indispensable subject, as there are no biomechanics without neurophysiology.

Nanne Boekholt, ICREO Animal Osteopath and rehabilitation trainer for horses at De Dierenosteopaat says it aptly:

When we are talking about coordination and balance in the body, but also about tension and relaxation, we are talking about neurology. The muscles, fasciae, limbs, and all other structures that determine position and movement are only the 'operatives' of the nervous system.

7.2 WHAT HAPPENS TO THE BODY WHEN TRACKING?

Horse who are tracking and who are able to maintain their body posture have a very low to mid-low head and neck position. They have long and elongated cervical and thoracic regions, which allows for joint mobility, movement, and space, which assists healthy cellular activity.

In order to take this natural elongated posture, the dorsal muscles are in a state of active relaxation. The pectoral muscles show good functional movement in supporting the weight of the thorax and forelimbs during movement. When these dorsal and pectoral muscles are touched, they feel toned but still malleable, giving a softer feel. In comparison, the dorsal and pectoral muscles of tense horses give a harder feel. They are toned but not as easily malleable. Tense horses could also experience compressed cervical and thoracic regions (**Figs. 7.1–7.2**).

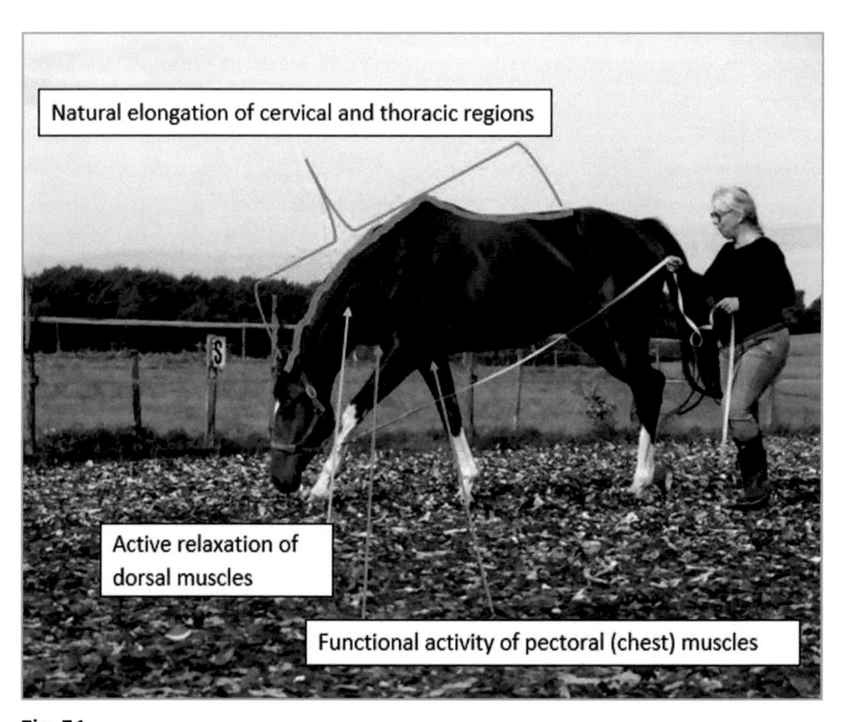

Fig. 7.1

Fig. 7.2

Horses who are able to maintain the tracking body posture will have dorsal interspinal ligaments and muscles that are in normal balance, supporting these normal and decompressed thoracic vertebra in the elongated regions. In addition, it is only possible to elongate the thoracic vertebra when the dorsal muscles, fascia, and ligaments are uninhibited and have good function (**Fig. 7.3**). Full flexion and relaxation of the back muscles enables greater mobility in the vertebrae. This mobility is almost impossible in horses who are in flight mode and tense.

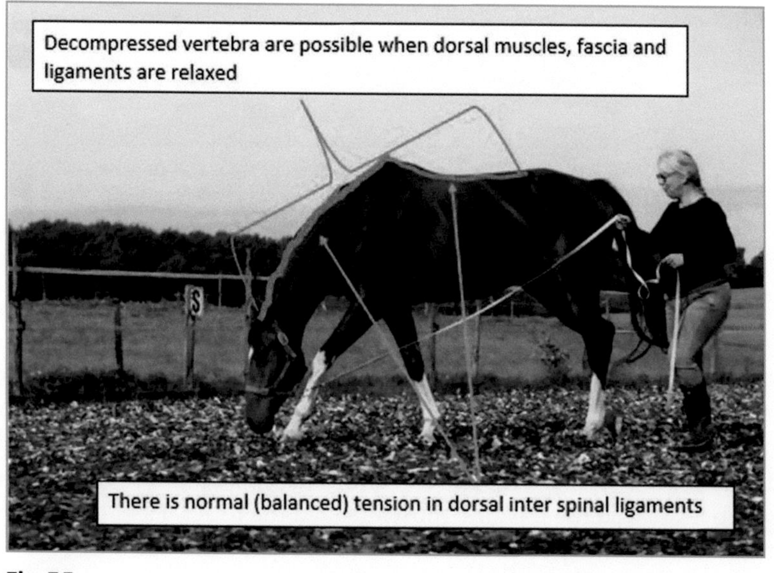

Fig. 7.3

When moving, the nuchal ligament and supraspinous ligament that run from the head to the sacrum of the horse play an important role by connecting and supporting all vertebra through the spinous processes. One of this structure's many roles is that it enables the fast elevation of the horse's head from a grazing or scenting position. When a horse's head is elevated, the contiguous ligaments (nuchal and supraspinous) are at their resting length, like a spring when it is coiled. However, when a horse's head is lowered, the ligament is like an opened, uncoiled spring, elongated and stretched. Therefore, when the horse lifts his head, it is like releasing an extended coil in a spring, allowing the horse to lift his head using less muscle power and energy.

From a muscular perspective, the naturally extended neck facilitates the activation of the gluteal group, and other deep hip stabilisers, that are critical for drive (both forward, breaking and reverse) as well as hip stability (**Fig. 7.4**).

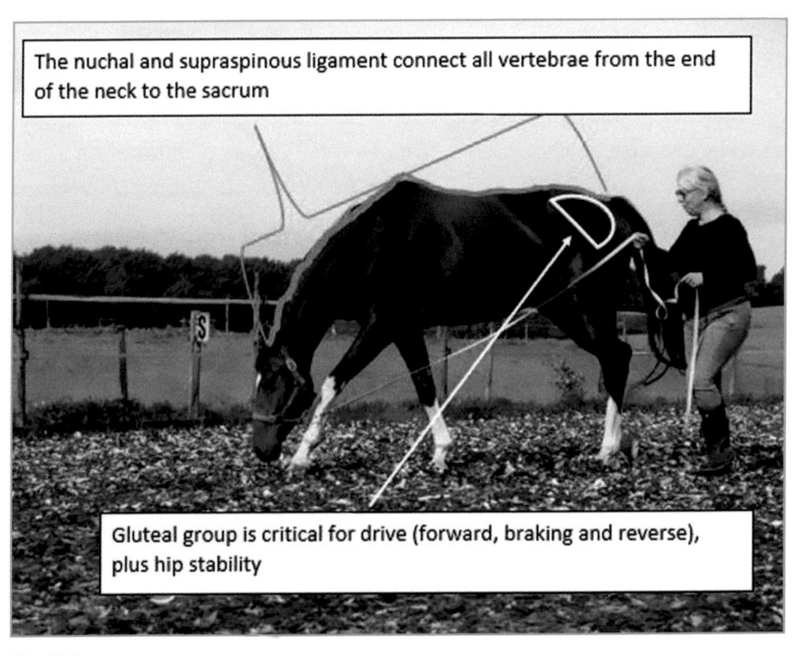

Fig. 7.4

When scenting, the horse's body is in a more natural position to support good functional movement. It enables the power muscles of the glutes and the hamstrings to extend the hip and, therefore, facilitate good 'drive' through the pelvic region. It also aids a natural extension through the hind legs, encouraging a lengthening of the deep hip flexor and the postural muscle, called the psoas. Healthy hip flexors will give more range and power to the drive of the horse through these muscles' ability to actively relax, enabling the extensor muscles of the glutes and hamstrings to extend the hind legs, allowing for a good range of movement.

Scenting also facilitates added freedom of movement for the forelimbs through aiding a decongestion of muscles in the neck and between the shoulder blades (scapulae). This allows the forelegs or limbs to move more freely, forwards (protraction), backwards (retraction), and sideways (adduction, which means toward the body, and abduction, which means away from the body). **Figs. 7.5–7.6**

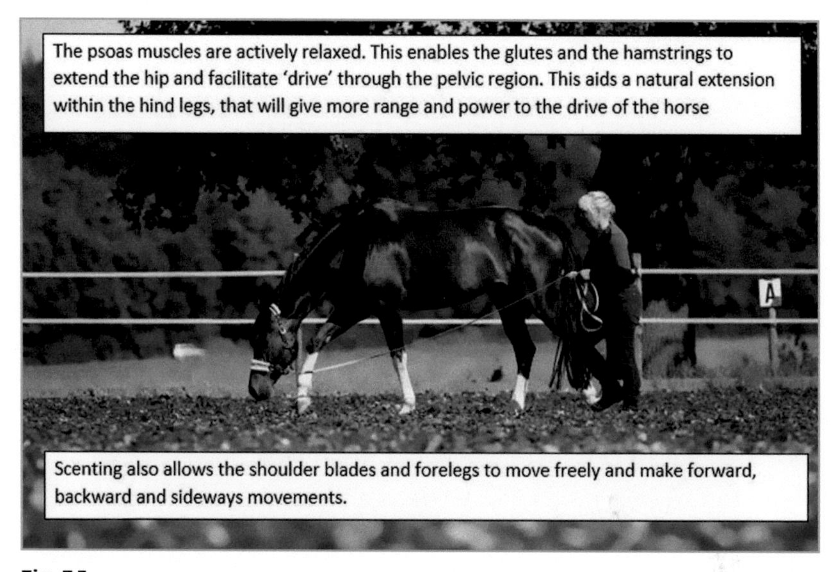

Fig. 7.5

Fig. 7.6

With natural scenting motivation, the horse will also gently abduct and adduct all four limbs, enabling natural development of the abductor and adductor muscle groups, pectorals/rhomboideus thoracic region and abductor and gluteal pelvic region, as well as other, deeper muscles that provide stability of the limbs, which is vital for good movement and biomechanics (**Fig. 7.7**). This, together with healthier epaxial muscles (dorsal interspinal muscles) and engagement of the hypaxial muscles (ventral muscles of the abdomen) gives a core engagement and stimulates core strength. This core engagement can be seen very clearly in the abdominal lift demonstrated in the **Fig. 7.8**.

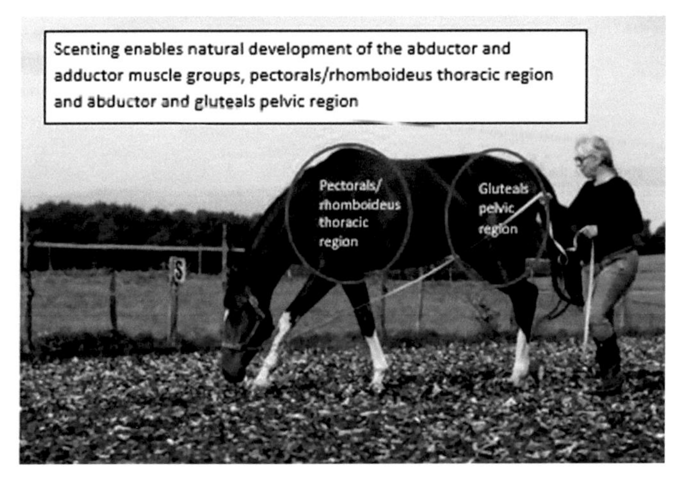

Fig. 7.7

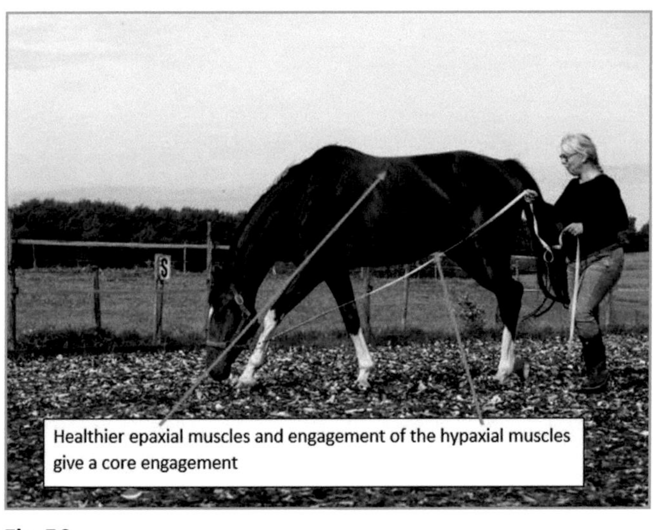

Fig. 7.8

A WORD ON BALANCE

When a horse is being ridden, there is a natural mechanism that puts the centre of gravity onto the forehand. The horse has to learn to also put weight on his hindquarters, balancing him out horizontally. Tracking, while being ridden or not, is a good additional exercise for this (**Fig. 7.9**).

Fig. 7.9

Vertical balance is about the horse dividing weight equally between his left and right flanks, and, for instance, not leaning when he makes a turn. Tracking is a good exercise for this. The lowered head facilitates a potential positive impact on the back. This, in turn, facilitates and encourages one of the many natural 'elastic' ligament and tendon mechanisms of the horse (**Figs. 7.10** and **7.11**).

Note: to give the horse a fair chance to practice this balance and coordination, we start tracking by the hand before we move to tracking beneath the saddle.

Fig. 7.10

Fig. 7.11

7.3 THE DIFFERENCES BETWEEN TRACKING AND GRAZING

Here, you could think to yourself, 'but don't horses have the same posture when they are grazing at pasture?' Nothing could be further from the truth. The posture taken while tracking at a walk is very different from the posture taken by a horse who is doing a bite-step-bite-step walk at pasture as he grazes. In lectures, I show a piece of footage in slow motion. Participants notice the following differences in it:

- The horses who are tracking walk at a higher pace than the horses who are grazing.
- They also move more quickly when turning.
- The hind legs step deeply under the body. There is optimal forward hip flexion with stretched stifle and fetlock. The fulcrum of the movement is at the hind leg and hip.
- The shoulder blades and forelegs are free to make forward, backward, and sideways movements.
- There is an optimal horizontal and vertical balance.

7.4 BIOMECHANICAL ADVANTAGES FOR THE HORSE

'Giving horses the opportunity to use their scenting ability, restores natural behaviours and anatomical functional movement. It also bestows autonomy, facilitating positive confidence through freedom of movement and thought.'

Director and Founder of Galen Myotherapy Julia Robertson

The biomechanical advantages of tracking for the horse are:

- Elongation of the body, allowing good flow of physical movement through the entire body.
- Natural elongation of the vertebrae.
- Improved muscle activation.
- Improved flexibility and agility of his body and movements.
- Improved posture and carrying capacity of his body, creating good self-carriage.
- Improved horizontal and vertical balance.
- Improved balance when tracking in different terrains outside arenas and paddocks.
- Improved balance between energy and relaxation while working.

SCENT TRACKING AS A SUPPORTIVE TOOL FOR HORSES WHO ARE RIDDEN

When horses are tracking and are able to maintain their body posture, they engage in a natural functional movement that motivates a dynamic process that requires the body to connect and work in a functional 'chain', linking muscles and muscles groups, fascia, to convert the bones and ligaments into moving levers, enabling functional movement that flows through the whole body, whatever the speed or gait.

As seen in the above examples and pictures, scent tracking can be very beneficial for horses. I also believe it may hold special value for horses that are being ridden. Although there are many angles to choose from, I want to share a few advantages scent tracking has for ridden horses that are directly related to the beforementioned biomechanical information.

When being ridden, some horses resist the load of the rider by holding the muscles in their back in a shortened position, which in time may have the effect of reducing the natural spaces between joints in the cervical and especially the thoracic vertebra. When horses adopt the tracking body posture correctly, they are naturally extending their heads and moving freely. The body is stimulated in a very natural way to elongate the cervical and thoracic regions. This may potentially reverse the loads and stresses placed on the vertebra created when being ridden. It may also stimulate and restore the normal balance within the vertebrae.

It is easier to elongate the thoracic vertebra when the dorsal muscles, fascia, and ligaments are uninhibited and have good function, but encouraging this movement through a natural activity could aid better function. Scenting stimulates that, as well as full flexion and relaxation of the back muscles, which enables greater mobility in the vertebrae, which is also needed when riding.

Ridden horses may experience fatigue of the muscles when carrying the load of the rider and executing riding programmes. Scenting without being ridden allows the hamstrings, psoas, and gluteal muscles to work without the loads and stresses of being ridden. This may have a different loading effect on the fibres, which could allow for positive cellular change.

Scenting also facilitates added freedom of movement for the forelimbs by aiding a decongestion of muscles in the neck and between the shoulder blades (scapulae), allowing the forelegs or limbs to move forward and sideways more freely. When the horse is scenting without being ridden, this freedom for the forelimbs is helped through the release of the areas normally covered by the saddle.

Scent tracking is therefore a great natural tool to restore anatomically functional movements, which is very useful for horses that are being ridden. Some examples for a comparison can be seen in **Figs. 7.12, 7.13, 7.14**, and **7.15**, and **7.16**

Fig. 7.12 Because of the optimal tracking posture, there is maximum freedom for the shoulder blades and forelegs to move freely and make forward, backward, and sideways movements. This is what you also want while riding.

Fig. 7.13

Fig. 7.14 During tracking, you have optimal forward hip flexion with stretched stifle and fetlock. 'Drive' is facilitated through the pelvic region.

Figs. 7.15–7.16 Tracking stimulates an optimal hip bend in the horse. The fulcrum of the movement is on the hind leg and in the hip, so the horse does not lean on his forehand. This should also be a goal in dressage. Muscle memory plays a part in this.

7.5 CONTRAINDICATIONS WITH REGARD TO BIOMECHANICS

I have not yet made any connections between scent tracking and the work of other professionals. Of course, it would be great if this were to happen in the future. If we had footage of horses with various ailments, we could perhaps create a tailor-made tracking programme for them and see if biomechanical improvements occur, and if so, which ones. It would really be interesting to be able to do this at a clinic. That way, the research could also be enriched with the ideas from the first part of this book. However, scentwork is used extensively within the canine world with Galen Myotherapy for dogs of all ages with conditions affecting mobility. Scenting is one of their tools to stimulate the body back to their natural function and natural biomechanical chains.

As I have indicated before, the horses that my customers and I track with are healthy. If there is some elevated muscle tension and stiffness, it is to the degree that the horse's body has adapted itself after only a few tracking sessions. Before this has happened, you could have seen that the horse was unable to stretch his neck down deeply and that he did not stretch his neck and back. You could also have seen that the horse walked a bit more stiffly, did not move his hind legs as deeply beneath his body when he walked, raised his head and neck a bit more often during tracking, or that he taught himself in the beginning to track without having to lower his head too deeply. A horse with poor muscle function can be seen to be standing still more often or, to the contrary, walking faster in order to maintain his balance (which makes the tracking more difficult). This could mean that when I started scent tracking with these horses, their muscles were inhibited or not functioning well. The gentle, natural scenting movement can begin to help ease these.

I have not tracked with enough horses who have metabolic problems and, therefore, are heavier and have stiff necks. There was one horse who developed a metabolic illness during the time he and his owner were tracking with me. When this happened, after a period of intensive treatment, the owner continued the tracking with a modified reward (see section 12.1). Another rare case for me was a horse who had the hiccups a lot. Tracking with this horse turned out to be the solution for stopping the hiccups on the spot. However, these are too few cases on which to base any conclusions. Other reasons why a horse might have trouble tracking are listed in section 6.3.

8 *Background information*

8.1 THE EXPLORING PHASE FOR SCENT TRACKING AND THE DIFFERENCES BETWEEN DOGS AND HORSES

A lot of horse people are also dog people. They share their lives and undertake activities with both horses and dogs. Many dog owners have done scent games with their dogs at different levels, from recreational to more professional. It is natural to want to copy the tracking methods you have used with dogs for horses. I did this too. I noticed, however, that there are essential differences between dogs and horses, and the results I was getting with various horses were too ambivalent, also in terms of the progress made. This is not something you want when you are trying to develop a methodology. After all, you are looking for a method that will appeal to every horse, that is feasible for every horse, and that motivates every horse to continue tracking because he enjoys it. Also, the method has to suit different kinds of handlers as well. I will discuss a few things I tried during this initial search and what happened.

WHAT HAVE I TRIED?

As I said before, I wanted to develop a method for teaching horses how to follow a scent trail, so it was crucial to find out what motivates a horse to do so. To do this, I made trails by dribbling small quantities of different substances to see if this would stimulate horses to follow them, such as applesauce, peach jam, or carrot pulp, as well as items such as bologna and salami. I tried different aromas from a bottle, such as lavender, vanilla, and pine scent, all of them with different degrees of dilution. I also laid different dung trails. These could be the droppings of horses familiar to the tracking horse (they went to pasture together, for instance), or of horses he did not know. The droppings also differed in freshness (in case you're gagging right now: I used gloves to lay the trail, putting down small pieces of excrement, and I also used pantyhose that I would put the droppings into so that I could drag them across the ground). I also tried urine.

I found that almost all the horses were not interested in the droppings of horses they knew. They probably knew this scent from the pasture they shared; however. the majority of horses were interested in the manure trail of horses they did not know. However, on a second trail, that manure could no longer hold their interest. They already knew it. The aromatic scents generally gained the horses' interest if they were diluted. If the scents were too strong, the horses showed evasive reactions. Salami and bologna inspired mixed results, from rare curiosity to evasion.

Using the scents (also in diluted forms) that had inspired interest, I began laying trails with enormous rewards at the end of them. However, I saw that many of the horses quit after about seven trails, divided across several sessions and days.

LESS IS MORE

I came to the conclusion that most substances give off a scent that is too strong, causing the horse to quit, either in the same session or after a couple of sessions. When I worked only with the scent of footsteps, however, which is very weak compared to the substances I experimented with, the horses did not quit. I cannot say what caused this for certain, but I did make two assumptions.

I think the stronger scents overstimulate the olfactory system, causing the horse to not inhale too deeply, and even to want to avoid the scent (and the ever-changing jackpot reward did not alter this fact). Incidentally, we know what this feels like ourselves when we are sitting next to someone who is wearing too much perfume. This scent is nice when it is applied lightly, but it becomes unbearable when someone is surrounded by thick clouds of it that we have to inhale.

In the case of a weaker scent trail created by footsteps, the horse has to use his olfactory abilities more intensely than in the case of a stronger scent. I can imagine that, because the olfactory system is used more strenuously, a proportionate quantity of dopamine is released (because multiple dopamine receptors in the olfactory tubercle are activated). This can then inspire more 'feel good' sensations than a trail with strong scents.

In light of this, it would be interesting to study the relationship between the intensity of the scent as well as scent processing and the quantity of dopamine that is released into the body. In my experience, this does not need to have anything to do with the length and duration of the trail. I think there is a correlation between the amount of effort it takes a horse to follow a trail and the wellbeing they experience from it.

FROM SIGHT TO SMELL

When I taught my dogs to do scent tracking, I used the method I learned at Anne Lill Kvam's Nosework Education. This method is also described in her book *The Canine Kingdom of Scent*. In this method, a pancake or a piece of sausage is tied onto a piece of string. A person lays a trail by dragging the piece of pancake or sausage across the ground away from the dog, moving out of sight of the dog as the trail is set. So, the dog has seen the person creating the trail, the walking pattern, the pancake, and the first part of the trail being laid but not the last small distance. If the dog wants to, he can follow the person laying the trail and get his reward. The method is expanded from this point. In my experience, this is a very good way to teach dogs how to track. That is why I initially also tried this method with horses who had never tracked before (without the pancake, of course). I had a reward the horse would really like on a rope, a person to lay the trail, and someone to momentarily gently hold the horse back (if he wanted to take off even though the trail was not complete yet, he was allowed to go). What major differences did I encounter?

First, that in many facilities for horses, it is difficult to find something you can easily hide behind (pastures and arenas are often open and flat). Not every horse owner has easy access to forests and bushes. With a dog, it is easier to go to such places.

Second, it seemed as if it is easier for dogs to switch between smell and sight. The horses were fine following the apple or other reward by sight, staying relaxed as they did so, but once they could no longer see it, they did not automatically switch to using their sense of smell to detect it (their facial expressions stayed the same: You would not see the nostrils expanding or contracting in any way in order to take in more scent).

I do want to stress here that I am talking about the first time the horses and dogs were confronted with this method and the degree to which, in a relaxed state, they switched between their senses of smell and sight during this first exercise; a horse who wants to investigate a stimulus or situation outside of this exercise will easily use both his sight and smell at the same time.

ACTION VERSUS CAUTION

I also saw a big difference in how horses and dogs approached new stimuli while they were searching. A dog will more easily and quickly approach and investigate objects that are new to him than a horse will. Horses are quicker to show a stop and flight response than dogs when they see, hear, and smell stimuli they do not know.

GETTING STARTED TOOK LONGER FOR SOME HORSES THAN IT DID FOR DOGS

In my experience, a greater percentage of horses than dogs had trouble getting started on the tracking. Many reasons I name throughout this book could play a role in this.

NOT WANTING TO FIND THE OWNER

Another big difference between these species is seen once the animals find their owners. With dogs, I have always seen intense joy when they found their owners again during sessions. With many horses I have seen, I do not see the same eagerness and joy. Some horses even turn away from their owners when they see them.

GOING TO ALL THIS TROUBLE FOR WHAT?

Another difference is the effort these two species want to make for their owners. I have a sense that dogs have a much greater will to please than horses do. Dog training programmes use varying quantities of rewards in order to enhance the dog's internal motivation. With horses, I have seen that this does not work. Regardless of the horse's aptitude, the reward has to be big from the very beginning, even for short trails. If the horse follows what we view as a very short trail and gets a small reward for it, it is hugely demotivating for the horse in the beginning. In dogs, however, I have the feeling you can get away with it.

DIFFERENT SURFACES AND WEATHER CONDITIONS

Of course, I also investigated how different surfaces and weather conditions impact tracking with horses, as well as what horses find easier and more difficult. From this emerged a method that was workable for me.

9 *The advantages of scent tracking*

When you regularly track with your horse, it has a great many advantages. The list in section 1.3 also applies to tracking, with a few small differences. These mostly flow from the fact that, in exploration and empowerment, multiple objects can be investigated with the sense of sight, smell, sound, and touch, and tracking mostly involves a single activity: the turning on and intense use of the nose. Of course, the horse can look at, hear, and feel environmental factors while he is tracking. For instance, he can walk and stand on different surfaces, step over obstacles, hear ambient noise, and look around at his surroundings. However, discovering objects is not as central to tracking as it is to exploration and empowerment. Tracking does have a few specific benefits though that are of great value. After reading about these benefits, you will understand why I am so enthusiastic about combining both disciplines.

9.1 HEALTHY HOMEOSTASIS IS STIMULATED AND MAINTAINED

In case you skipped Part 1 because you were so enthusiastic about tracking, you can find more extensive background information for the points I will address in the upcoming paragraphs in Chapter 2.

In section 2.3, I explained that exploration and empowerment is the instrument for stimulating and achieving healthy homeostasis: the state in which the body functions optimally and is able to deal with changes without these causing additional compensations or tension. This is just as true, if not more so, for tracking, particularly because the olfactory system and the ability to smell for long periods is stimulated.

Among other things, scent tracking stimulates the olfactory tubercle, which is densely wired with dopamine receptors. By stimulating the horse to smell more intensively, we activate the release of dopamine into the horse's system. Dopamine is a neurotransmitter that is directly involved in the feeling of reward and wellbeing. It plays a strong motivational role and spurs goal-directed behaviour. Dopamine is almost addictive.

Horses who deal with many moments of acute stress and chronic tension and horses who deal with depression have the same underlying biological problem: They both have decreased dopamine levels. In addition, a horse who lives with a lot of moments of acute and chronic stress also has an elevated cortisol level. Letting the horse discover, have new experiences, and solve problems (what I think tracking is) stimulates the release of dopamine. This helps the horse to achieve healthier homeostasis because it gives low dopamine levels a boost. It also helps the horse who lives with a lot of moments of acute or chronic stress because the activation of dopaminergic pathways causes an inhibition of the release of corticotrophin-releasing hormone. This also reduces the release of cortisol.

Just as described in section 2.3, brain plasticity is also stimulated through tracking. The dendrites acquire more bifurcations, and the axon terminals grow new sprouts that could grow in a new direction. This is beneficial because the more neural connections a horse has, the more he can use them to understand and remember places and things. In addition, the neural connections that are used often are myelinated more, improving the electrical conductivity to other cells. New cells are created in the hippocampus, the place where memory is stored. This plays an important role in thoughts and actions. Thanks to stored memories, a connection can be made between the thoughts and experiences of the past and the ones happening now.

9.2 SCENT TRACKING IS POSSIBLE IN A STIMULUS-POOR ENVIRONMENT

What is very nice about scent tracking with horses is that you can also do it in a stimulus-poor environment. You can do it in an arena, pasture, or paddock at the facility where your horse is housed. This is an advantage when you are working with horses who are still learning to get used to all sorts of stimuli in their (new) environment that might still cause them tension, or horses who cannot handle much for other reasons, such as they are or have been in pain or are recovering from tension or depression.

Tracking might help horses who are still experiencing tension to gain or maintain a better balance. They could also be horses who have just been moved to a different stable, have to stay in veterinary clinics, are going away on competition, or have to stay in a different location for a few weeks. It could also involve horses who cannot freely go to pasture for some reason, but who you do want to give a chance to show their natural behaviour in order to give them the benefits of tracking, exploration, and empowerment. If you have horses who are capable of handling only very few stimuli, I advise starting the tracking in a place the horse is very familiar with and that has few stimuli. If you have been tracking for a certain time frame (depending on the tension level of your horse), you can slowly start doing exploration and empowerment exercises. You might then start introducing regular training exercises.

But, as I said before, scent tracking, exploration, and empowerment are suitable for every horse, including the mentally and physically healthy horse who is perfectly in homeostasis. Why? Because horses love doing it so much, and because it is mentally and physically good for them.

9.3 HORSES LOVE SCENT TRACKING

Once they get the hang of it, they cannot wait to do it again. If they are at pasture, they will run toward the fence when they see me or their handler with the yellow tracking rope. They sometimes trot to the starting point of the trail; that is how much they want to get started. And horses who no longer want to work with people get back into a spirit of wanting to do this again. Tracking returns a bit of individuality to the horse. You let him show functional behaviour that had been neglected in his domesticated life.

9.4 THE ROLE SWITCH BETWEEN THE HORSE AND THE HANDLER

The joy of exploration and ultimately finding is multiplied when it's a shared exprience between horse and owner.

Co founder of the field of Animal Assisted Play Therapy and Co director of Turn About Pegasus Tracie Faa-Thompson

A friend; a sympathetic ear; an outlet; a means of transport; a living exercise machine; a lawnmower; a mirror; a guide; someone to vent your frustrations on; a way to get attention, to generate money, to gain status, to generate fun and happiness, to…what else?

The horse plays many roles for people, roles that all have one thing in common: they serve *us*. He has to conform to what we want; we use an entire arsenal of techniques to make it so and keep it that way. No matter how legitimate we think this is and what arguments we use, it is good to realise that many of the tasks our domesticated horses perform are far removed from the lives of his semiferal or feral conspecifics. That is not meant to idealise the feral existence, but to point out that what we are asking of the horse is not nothing, especially when it comes to conformity and obedience.

One of the best things about tracking is that there is a role reversal in terms of control, in particular between the person and the horse. Not only that, but it happens during an activity he really enjoys (or will start to like after doing it a few times). Now, the horse can do something the person cannot. The person can only follow and hope the horse gets the job done. And if the scent bag or missing person is found, the person helps the horse to open the bag and hold it up while he eats.

That shifting dynamic is very special. The role reversal of the person following the horse for once instead of the other way around, can inspire many different feelings and behaviours in handlers: from light to high tension, insecurity, frustration, and boredom (if their horse is taking too long), to great joy, amazement, connection, and pride, especially when the horse completes the task. You open the bag for him, and you enjoy the victory together as he eats. In his book *Behave: The Biology of Humans at Our Best and Worst*, Sapolsky referred to a test in which two test subjects were playing an economic game. If they worked together, they both got a moderate reward. If one excluded the other from the game entirely, he got everything, and the other person got nothing. The result was that test subjects who both chose to get a moderate reward had a higher dopamine release than the subjects who made the selfish choice.[1] So, working together seems to cause an added release of dopamine. I can imagine that this might also be true for a horse and handler in the case of a successful tracking run.

As you can see, tracking can bring out a lot of feelings and create many learning experiences when you are teaching your horse how to do it or when you are working with a horse who already knows how to track. It is about the excitement, choosing the ingredients for the scent bag, figuring out what your horse really likes, burying the bag, doing your best not to give away its location, collecting the horse, hoping he will want to search for the bag (you also have to hold back so as not to distract the horse: see the practical chapters), and then experiencing what it is like when you relinquish control and follow the horse. You might also experience disappointment if the horse does not find the bag (at which point you will try to figure out your next move). But maybe the horse will find the bag, and you will share the joy of this together.

As tracking with horses requires coming up with and executing the practical steps, thinking about the other, letting in feelings and different experiences, seeing the other and taking him into account, playing and sharing together, you can see why such an exercise might be a good discipline to include in animal-assisted therapies. Together with Dr. Risë VanFleet and Tracie Faa-Thompson (www.iiaapt.org), the women who cofounded the field of Animal Assisted Play Therapy®, we made a start at developing such a programme. They both immediately saw the relevance of this work to their own and made arrangements for the professionals in training in their programme to learn more about it.

CONNECTEDNESS AND RECIPROCITY

How the horse handles this role reversal depends on his nature, how he is feeling mentally and physically at the time, and his life experiences. You can have a horse who immediately makes decisive and proactive use of the freedom he is given. He assertively gets to work, but there are also horses who have to get used to their newfound freedom and who only make use of it bit by bit. What is true of all horses, however, is that you will see them flourishing once they take charge in the tracking task and complete it successfully.

Another wonderful side effect that others and I have experienced is that the more we let horses make their own choices in exploration and empowerment exercises or tracking, the more easily they will acquiesce when we ask something of them. In turns out there is some sort of credit and reciprocity going on that does not just exist between humans, but that horses also understand well and apply naturally. Perhaps this happens because they start to feel better because of the exploration and empowerment and tracking exercises, or because they start to feel more connected to us when we track with them. After all, eating together, taking the horse for a walk on which he can graze, and searching for food together are quintessential bonding activities. I often hear people say that they feel connected to the horse when they are riding or touching him. As an extension of this, the tracking rope can be an additional instrument for achieving that feeling of connection with the horse and tracking process, of course, with a smile in the rope.

A SUMMARY OVERVIEW OF THE BENEFITS OF TRACKING

For the sake of convenience, I will list the benefits of tracking:

- Tracking meets the horse's natural need to search and explore.
- Tracking meets your horse's need for independence and freedom of choice.
- Tracking boosts joy in life and eagerness to undertake activities.
- Tracking increases the inclination to try new things and investigate further, including unfamiliar objects and locations.
- Because the horse is not influenced by others when he is tracking, the focus is entirely on him.
- While he is tracking, the horse experiences the environment. He gets to know it and can place it within his own frame of reference. Because there is no exterior compulsion, there is a good chance that he will create positive associations with these environmental stimuli.
- Because the horse is in control of his own tracking pace, there is a better chance that he will stay within the tension zone he can handle. If he is unable to do this, this method allows him to practice with this on his own. In case of hesitation, he can overcome barriers and practice with self-regulation.
- Because the horse is in control of his own tracking pace, the rider or handler gets valuable information about the horse.
- Because the horse gets to explore in his own way and is in control of his own pace of exploration, raising the chance that he will remain inside the tension zone he can handle, there is a good chance that the information will be stored in his long-term memory.
- Tracking improves his body posture and gross- and fine-motor skills.
- Tracking calms him and tires him mentally and physically. The more intense the tracking, the bigger this effect will be.

If you do it regularly (for instance, twice a week on average), tracking causes:

- A decrease in fear and aggression responses with regard to known stimuli.
- A decrease in fear and aggression responses with regard to new stimuli.
- A decrease in impulsivity.

- A decrease in overreactions and tension.
- A decrease in frustration (which had been caused by not being able to execute natural seeking behaviour).
- A decrease in boredom and tension caused by boredom.
- A decrease in the development of chronically elevated stress levels.
- A decrease in the chance of developing 'shutdowns', learned helplessness, depression, and lethargy.
- A healthy immune system and a lower chance at developing stress-related ailments.
- An increase in impulse control.
- An increase in the amount of time in which a horse can concentrate.
- An increase in problem-solving ability, or an increase in displaying problem-solving ability.
- An increase in long-term memory capacity.
- An increase in happiness and joy in life.
- An increase of calm in the horse's behaviour and nature.
- Better body control and a more flexible body.
- An increase in seeking out and maintaining social relationships.
- An increase in engagement with the rider and/or handler, as it improves the relationship between horse and human from the horse's perspective.
- An increase in comfort with regard to the presence of other people.
- An increase in willingness when it comes to the tasks people ask him to perform because of the development of reciprocity.

Fig. 9.1

REFERENCE

1. Sapolsky R.M. (2017) *Behave: The biology of humans at our best and worst*. Penguin Random House, 66.

10 *Practicalities: What you need to know*

10.1 WHAT SCENTS DOES A HORSE FOLLOW?

Let's get into the practical side. Here you will find useful information for when you want to start trying things out. We will start with the scent your horse will be following.

SCENTS THAT EMANATE FROM THE PERSON LAYING THE TRAIL

Out of all the scents a horse tracks, one of them is the scent of the person laying the trail. No matter how well you have washed yourself, we all carry an individual odour. This scent is a blend of the scents we excrete from our pores as a consequence of what we have eaten, such as garlic and spices, and scent particles we shed while moving, such as skin flakes, skin cells, hair, the fragrances of lotions and perfumes we are wearing, or miniscule clothing particles. Without rain or wind, these odours will linger on or above a trail for about half an hour.

SCENTS THAT EMANATE FROM THE SHOE ITSELF

Before I started tracking, I had never stopped to think about the soles of my shoes. Sure, when they became worn, I got the shoes resoled, but that was it. Now that I do tracking, this has changed. First of all, I have become aware that the sole is worn out by touching different surfaces and that the wear and tear is caused by sole particles tearing off and landing on the surface. The horse can also smell these torn-off particles. Second, I think about the type of sole the shoe has. The more grooves it contains, the more easily all sorts of matter sticks to it. These miniscule waste particles are also deposited on the trail (**Fig. 10.1**).

Fig. 10.1

As such, it is good to think about the shoes you wear when walking a trail. You can choose to help your horse with his first trails (or always) by wearing shoes with deep grooves in the soles. If need be, you can trample around a pile of manure or a dirty stall with the shoes first. The miniscule scent particles, which are not necessarily visible, will then end up on the trail, and this can help the horse. However, if your horse is an experienced tracker, you can choose a smoother sole that sheds fewer scent particles.

Also interesting is information from the study "The Horse's Foot as Neurosensory Organ: How the Horse Perceives its Environment," by Robert M. Bowker, et al.[1] in which the researchers suggested that when horses set a foot down on the ground, the scent glands in the zone of the central sulcus or the caudal frog are activated, causing an apocrine-type secretion. So, it is easy to imagine that the moisture that is left behind with every step a horse takes is a good way for other horses to follow him at a distance. Our shoes seem meagre in comparison. Luckily, though, our scent also seeps out through our shoes (**Fig. 10.2**).

Fig. 10.2

A PERSON'S SCENT COMING OUT THROUGH THE SHOE

My son played football for years. He often left his closed sports bag containing his dirty uniform, wet towel, and muddy shoes in a corner of the hall and left there. After a day or so, however, we were forced to clean it up because the bag began to smell. The heat of the house and the humidity inside the bag allowed scents to develop optimally, especially the smell of butyric acid: a component of our sweat. This scent is sometimes described as that of dirty socks or vomit. Although there are many scents we cannot perceive, butyric acid clearly is not one of them.

Even if you shower every day and keep yourself clean, you still secrete the smell of butyric acid. And for all of us, it seeps out through our shoes. According to Anne Lill Kvam, it takes an average of 8 hours for butyric acid to leak through a one-centimetre sole. As such, a shoe that you wear for many hours and days is steeped in butyric acid, of which scent particles are deposited with every step you take. That is why it is easier for a horse to follow a trail if you wore your oldest, smelliest shoes to lay it, as they are drenched in your butyric acid scent, meaning you have left the most miniscule molecular particles of yourself on a trail.

DISTURBANCES OF THE GROUND AND LIVING ORGANISMS

Other scents your horse follows when he is tracking are those that are released when you disturb the top soil with your footsteps. The earth that is churned up emits an odour, as do the blades of grass you break, and the seeds you crush. You can also crush insects, such as beetles and ants, with your footsteps. This can happen above and below ground. If it happens below ground, it is because the earth itself is 'compressed so that little pockets under the ground are squeezed and small amounts of gas or liquid is released.'[2]

BACTERIA CLEANING UP DEAD AND DYING PLANTS

In nature, the crushed plants and insects are cleaned up by bacteria. These bacteria attach themselves to the insects, producing a whole new scent. A horse can also follow this scent if he is following an older trail. However, bacteria need water to start their digestive process. If it is hot and there is no water in the ground, these odours will not release as effectively.

TIP: When you are using the paragraphs above to teach your horse to follow a scent trail, wear shoes you have already worn for many hours (that way, your scent, and especially the butyric acid, has seeped through the sole and leaves a mark with every footstep) and shoes with deep grooves in the soles, as they disturb the ground well and might also deposit some scent particles that had been stuck to the soles.

Many of the horses I track with are interested in the soles of my shoes at frequent instances along the way. They sniff it, and when I lift my foot so they can smell the sole, they eagerly take advantage of it. (These are also the shoes I use to lay trails.)

10.2 AIR SCENTING

Through air scenting, the horse can also catch scents that are floating around in the air and that can travel significant distances on the wind. These scent particles become stronger as you approach their source and weaker as you retreat from it. Air scenting is something we humans do too, although perhaps less consciously. For instance, when I am on the top floor of a house in which someone has baked an apple pie, I can smell the scent floating through the air. If I want a piece, I can follow the ever-intensifying scent to its source downstairs.

Horses use air scenting a lot. It is an easy way to locate, discover, or recognize the source of a scent. If you pay attention, you will see that horses do this frequently. They air scent while being ridden, for instance, or when they are at pasture and someone passes by at 50 metres distance. You can see by their nostrils that they are breathing a bit more rhythmically and in a more pronounced way. Their head positions vary from mid-low to high.

While tracking, I often see horses holding their heads between the mid-low and mid-high positions. Also characteristic of air scenting is that the horse walks in fairly straight lines (**Fig. 10.3**).

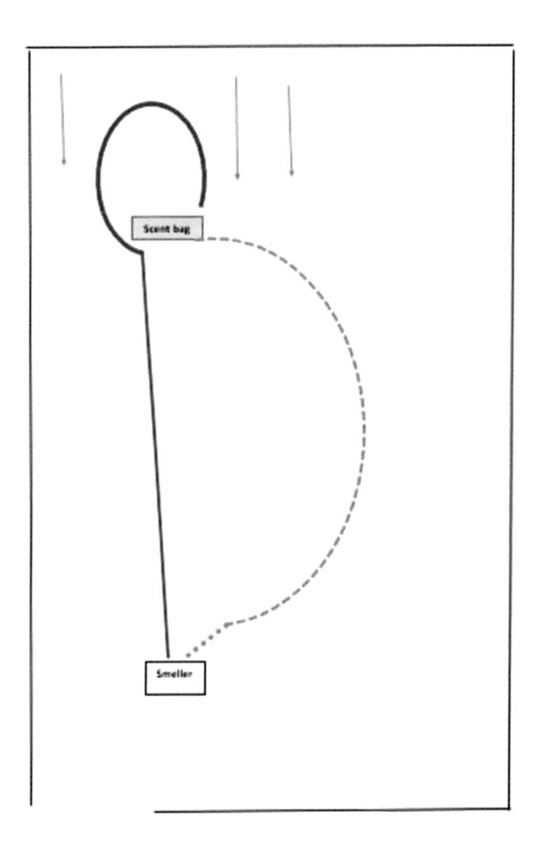

Fig. 10.3 The orange line reflects the route I walked: a clear, wide arc to the right. Indy does not follow this, however. When she arrives at the smeller, she walks to the bag in a straight line, her head position between horizontal and mid-high. The direction of the wind aided her in doing this. After arriving at the reward bag, she makes another lap just in case, once again ending up at the bag. Treasure found.

10.3 SCENTING ON DIFFERENT SURFACES AND IN DIFFERENT WEATHER CONDITIONS

SCENT TRAILING ON DIFFERENT SURFACES

Where your horse is stabled also influences the way you can lay a trail. Is there an outdoor or indoor arena there? Are you tracking on sand or grass? Is there an option of changing locations? Is it possible to lay long trails there, or not? Chapter 15 explains how to make the trail more challenging if you have a smaller or larger space at your disposal.

If you have the intention to go tracking with your horse in large areas and creating long trails, you can choose to do this in the large area in question from the very beginning, or you can choose to start in an arena. Then, once your horse learns how to track, you can start laying trails in different places.

When you start, this is what you what you should know about the different surfaces:

- Sand – This is generally an easy surface on which to start tracking. Sand buckles easily, allowing any insects to be killed and scent to be released. The churned sand itself can also release an odour. Human scent particles can also be smelled well on sand.
- Artificial floor covering – An arena with artificial floor covering, such as turf, can be more difficult for a horse. The material is often very soft and springy. It will be difficult for scents to be released from the sand below the covering, or from insects being crushed. The covering itself can also carry a scent. The horse can smell these particles very well. If your arena is covered with an artificial material, stamp your feet extra hard to release scents (mark these places clearly) or place your feet in between the coverings, directly onto the ground.
- Forest floor –This is an easy surface to work with. The scent particles the person laying the trail leaves are easy to smell, but the surface also damages well when the trail layer walks on it. Because there are many insects in the forest, there is also a better chance of crushing them as you walk, releasing scent. And the scent bubbles of compressed air can also rise well.
- Grass – A grass surface is well suited to tracking. Grass also retains the scent particles of the trail layer well and buckles easily, allowing for insects to be crushed, releasing scent. The difficulty with grass, however, is that it emits a strong odour when it is freshly mown and that a farmer might have sprayed it. Pesticides often also carry a strong scent. In addition, tracking on grass can be harder for horses who do not often encounter it: they might associate it first and foremost with a chance to graze. Horses that often graze on a grass pasture or always have roughage at their disposal do not have as much of a problem with this.
- Stone – It is more difficult for horses to follow trails laid on stone. Of course, it retains the scent particles of the trail layer, but stone does not buckle like earth does, and so there is no scent from the ground being churned up, insects being crushed, or air pockets being compressed.

TRACKING IN DIFFERENT WEATHER CONDITIONS

I have done most of my tracking in the Netherlands, a country that, according to the Köppen classification, has a temperate maritime climate. This means we have mild winters, mild summers, and rain all year long. In terms of weather, when do I find it hard to track, especially if the horse is an inexperienced tracker? When the rain is so heavy that trails are diluted and washed away. When the wind is blowing so hard that scents are dispersed. (This is approaching storm levels, however. Your horse can cope with lighter rain and a slightly stronger wind.) I have also noticed that the Dutch horses find it more difficult to track on very hot and dry days, when the temperature rises above 30° Celsius. It seems as if they are willing, but the trail just gives them less to hold on to. They will begin to follow it, but seem less able to stick with the trail. I can imagine that this also has something to do with the humidity. If I want to revive a scent, I can literally breathe on it. My warm, moist breath causes scent particles to revive so they can be smelled again. In the Netherlands, in hot temperatures, the humidity is low, so the scent is probably harder to smell. It is also possible that the lining of the inside of the nose is dry. And there is probably also a familiarity factor: Dutch horses are used to tracking in lower temperatures. Anne-Lill Kvam told me that tracking with dogs is more difficult below –18° Celsius because the butyric acid freezes below that temperature. I have never tracked at these low temperatures simply because I have never been to such a cold place.

REFERENCES

1. Bowker R.M., et al. (2012) *The Horse's foot as a neurosensory organ: How the horse perceives its environment*. p. 26. https://www.researchgate.net/publication/305766078_The_Horse's_foot_as_a_Neurosensory_Organ_How:the_Horse_Perceives_its_Environment
2. Kvam A.L. (2012) *The canine kingdom of scent*. Dogwise Publishing, p. 77.

11 *The four pillars of this method*

THE FOUR pillars of this method include:

1. From the start, the horse works only with his nose.
2. Seeing the cloth (the smeller), means that scenting has begun and there is a reward to be had.
3. Association with the footsteps on the cloth helps the horse to reach the reward.
4. If your horse is an experienced tracker when it comes to finding the bag, you can move to tracking missing persons.

PILLAR 1: FROM THE START, THE HORSE WORKS ONLY WITH HIS NOSE

My tracking method is characterised by a return to basics by imagining what suits the nature of the horse, i.e., how he would be if his life were not dominated by humans. Looking for food would play a prominent role in that case. It is a natural behaviour that is part of the horse's ethogram.

Although looking for food is natural behaviour, there is a percentage of horses who do not naturally use their noses. It is as if they have forgotten that they can 'switch them on'. They do a quick search with their eyes, and if they do not see anything, they stop looking altogether. Many horses disengage when they do not see the thing they are looking for. Not only that, but the next time they are far less motivated to search at all. However, when I had such horses use only their noses from Step 1, they were motivated to search the next time, too, even though Step 1 could be difficult, and some horses did not grasp the tracking right away. I think this has something to do with the fact that only very little dopamine is released while looking at things, while a lot more is released during tracking (see section 6.2). Stimulating and using this dopamine secretion is part of my underlying vision. That way, the horse feels a difference in his body right away, and he is motivated to repeat the experience the next time.

As such, the first pillar of my tracking method is stimulating the nose, not the eyes, from the beginning. That is why I have developed a bag that you bury, thus removing it from the horse's sight. The bag has a section made of mesh, the structure of which is loose enough to allow plenty of scent to pass through to stimulate the horse, but not so loose as to allow sand or dirt to trickle in. The bag has been sown in such a way that it is very flat, enabling you to cover most of it up, leaving only the part with the mesh visible above the surface. It is also easy to fold the bag, so that you only have to bury as small an object as you need to (**Fig. 11.1**).

Fig. 11.1

PILLAR 2: SEEING THE CLOTH, THE SMELLER, MEANS STARTING THE SCENTING. THERE IS A REWARD TO BE HAD

The scenting starts with the smeller, the piece of cloth that forms the start of the trail. Use of the smeller teaches the horse that he needs to follow the scent on the smeller, not just any scent trail. Of course, it is not a bad thing if he does follow other trails. Horses want to do this. It can sometimes be difficult for riders and handlers who have a horse who constantly lowers his nose to the ground. That is why we make this distinction that is understandable for the horse (**Figs. 11.2–11.3**).

Fig. 11.2

Fig. 11.3

PILLAR 3: ASSOCIATION WITH THE FOOTSTEPS ON THE CLOTH HELPS THE HORSE TO GET THERE

The horse learns that there is a reward to be had when he sees the smeller. In addition, after a few sessions, he learns that the scent of the footsteps that are on the smeller will also lead him to the bag, if he follows it (**Fig. 11.4** and **Fig. 11.5**).

Fig. 11.4

Fig. 11.5

PILLAR 4: THE HORSE FIRST LEARNS TO FOLLOW THE SCENT TRAIL TO THE BAG, AND THEN YOU MOVE TO TRACKING MISSING PERSONS

A more detailed discussion of Pillar 4 is provided in Chapter 17.

12 *What materials do you need?*

To track, you need a reward, a scent bag, a smeller, a long lead rope, a halter, and maybe a small shovel (if you are searching for missing persons, it goes without saying that you also need a missing person).

12.1 THE FOOD REWARD

Certain scent molecules are picked up more quickly and are stronger than scent molecules that are heavier. An example of the latter is the banana: the scent molecule amyl acetate is more difficult to dissolve and, therefore, more difficult to smell. If you have feed, horse treats, or small bits of food, you can put these into the bag as well. Some of these carry a strong scent, which you can use if needed (**Fig. 12.1**).

Fig. 12.1

It has generally been my experience that recently cut fruits and vegetables that are wet are easily smelled by horses. If you use apples, feed them to the horse without the cores. Apple seeds contain arsenic, which is poisonous; you do not want your horse to ingest too much of this. You can also think of other types of fruit, such as apricots or peaches (without pits), watermelon (without the rind and seeds), mango (without the pit), pears, strawberries, or raspberries. It is important that your horse sees it as a reward, so try it out first, that way you will know which foods he *really* wants to find (they then act as a reinforcer). It helps if the treat also carries a strong scent.

Be aware that fruits and vegetables contain a lot of hidden sugars. If, for health reasons, your horse is not allowed to ingest sugar, you can use rewards such as spinach, celery, or lettuce. If he is only allowed hay, use that. You could mix a few apple slices in with the hay, and then take them out again when you start. Hopefully this apple-flavoured hay has a good smell and flavour and will make the horse feel good rather than cheated.

MAKE YOUR REWARD LARGE AND VARIED

My scent bag does not contain just one reward. Its contents is varied, and I change it every tracking session, making different combinations of the foods the horse loves. That way, what the horse will find is always a surprise and the food reward does not function as an ordinary motivator, but as an *extra* motivator. It makes finding and eating it a real treat every time (**Fig. 12.2**).

Fig. 12.2

Do not be stingy with the quantity, even for short trails. Tracking is much more tiring than we think. It is hard work, and the reward must be proportional to the horse's effort. A good rule of thumb is that it should take your horse at least 3 minutes to eat everything in the bag once he has found it. If he licks the inside of empty bag, that is fine. Let him. You now know that you have found a good reward, and it is good that the reward moment lasts a long time. Next time, he will no doubt be willing to try hard again for such a long reward.

Tip: Cut your vegetables and/or fruits at home and bring them in a well-sealed container so that they do not dry out. Cut the fruits and vegetables in such a way that the horse can eat them easily and does not choke on them.

MAKE A TOP 10 LIST

Tip: Make a top 10 list of the things your horse really likes. You can do this by offering him various foods at the same time and seeing which he chooses to eat first. This can be grains, pellets, fruits, vegetables, etc (**Figs. 12.3–12.4**).

Fig. 12.3 We offered Hope several pieces of fruit and vegetables. On the cloth were pieces of pear, apple, watermelon, carrot, banana, and some strawberries. Hope got two separate turns to investigate and choose what she likes. She was quite precise; her choices of importance were first the pear, then the carrot, and then the apple. She sniffed quite intensely at the watermelon, banana, and strawberry but did not taste them.

Fig. 12.4 Vosje got two rounds of tasting as well. He did not seem to show any particular preferences, but methodically ate all the fruit and vegetables from left to right. However, he skipped the strawberries. So it could be the case that, apart from the strawberries, he loves all the foods that were offered equally; however, if I wanted to double check, I could also check his preferences by using two different pieces of food, holding one in each hand (for instance, a piece of apple in the left hand and a piece of watermelon in the right), closing my hands loosely, and holding them in front of Vosje. I could then give him the piece of food he seems to be the most interested in. I could repeat this with different kinds of foods.

12.2 THE SCENT BAG

When I started tracking and this method began to take shape, I began to develop an interest in creating a suitable scent bag, also because such a thing did not exist yet, and the many different bags that did exist did not lend themselves well to the task. For instance, small particles would detach from the interior, meaning the horse might ingest fabric fibres. The bag often was not made for the anatomy of a horse's head, sometimes making it difficult for the horse to breathe while eating. Feed mixed with saliva would stick to the seams, so the bags quickly became filthy, and they were hard to clean and dried badly. So, carrying a long list of desired features, I approached my son, Sybrand Jansen, who, at the time, was studying fashion design at the Utrecht School of the Arts. He made several prototypes, and for a few months we tried out various materials, sizes, types of stitching, zippers, and mesh. Tailor Paul Stam turned this prototype into a model that could be handmade in a relatively short time (**Figs. 12.5–12.6**). This work resulted in a bag that:

Fig. 12.5

Fig. 12.6

- Unfolds in such a way that there is enough room for the horse to breathe while eating.
- Is wide enough that it leaves enough room for the horse's mouth.
- Is easy to hold, and when it is spread open more widely, the shape holds and remains comfortable for the horse.

- Is constructed with the seams on the outside so that the interior is seamless.
- Is easy to clean; you simply rinse it out and hang it up to dry.
- Has a small window made of mesh, the openings of which are large enough for enough scent to pass through, but small enough not to let in sand, soil, leaves, etc.
- Uses fabric to which food does not stick.
- Has been sown in such a way that it can lay very flat so that you do not have to dig a deep hole to bury it fully. This is convenient, especially when the soil is hard.
- Bends easily, as does the zipper, meaning that if you bury it, you can do so in a smaller hole.
- Uses fabric in colours that blend nicely into the environment.

12.3 THE SMELLER

Fig. 12.7

The smeller is a piece of cloth that marks the starting point of the trail. For me, the ideal size for the cloth is 30 by 30, 30 by 40, or 30 by 50 centimetres. You can often buy these in stores under the name *guest towel*. It can also be a cleaning cloth. The piece of cloth is made of towel terry. The cloth's size makes it easy to fold and carry. It is also not so big that the horse has the association of wanting to stand on it (if he was taught this in the past).

I usually use the colours white, yellow, or bright blue. Studies show that horses are able to distinguish the colours in the yellow and blue palettes well. They appear to be less able to distinguish the colours in the red-orange palette. They tend towards green.[1]

12.4 A LONG LEAD ROPE

It is important that your rope is longer than the regular halter lead rope of 1.5 to 2 metres. You want to have enough room to walk by your horse's flank and hindquarters. Say, you are walking on his left, and he makes a sudden turn to the right, you would almost have to run to keep up with him, especially if he is moving fast because of the tracking. In this situation, you do not want to accidentally yank on the rope and hold back your horse. If your horse is an experienced tracker, a small tug will not slow or stop him, but in the early stages of teaching your horse how to track, he could interpret a tug as a correction or instruction for him to stop, making him abandon the trail. That would, of course, be a shame. Personally, I do not like to use ropes as long as lunging reins, because I do not like the feeling of constantly having to roll-up and mind the length of the rope. My ideal length for a rope is 4 metres.

The material I prefer is biothane. It is a little bit stiffer, which is easier to walk with. It does not swing as much when making turns. What I also like about this material is that is does not easily get caught on trees or bushes when you are tracking in the woods. It can be cleaned easily. Also, biothane is often available in various colours, which I find useful. That way, if I am tracking in public spaces with my horse, I can use a bright yellow or bright orange rope to make sure cyclists or pedestrians see it at a distance. Of course, you can make your own choices here.

It is important that the hook that attaches the rope to your horse's halter is light, and that there are no heavy braided sections beneath it because, when the horse is walking or trotting with his head low to the ground, a heavy hook will often drag across the ground. This is distracting, and it also increases the risk of the horse stepping on the rope. You will have to expend more effort to keep an eye on and lift the rope, which detracts from observing and enjoying your horse.

In addition, the weight of a heavy hook and rope are of no added value to the tracking. If you have a horse who might have been trained to obey with the same kind of weight or tugs on his halter, you will avoid any negative associations and learned behaviours associated if you forego the heavy hook. I prefer a small, light-weight connection for tracking.

WHY USE A LEAD ROPE AT ALL?

I am sometimes asked why you should use a lead rope at all. Why not just let your horse walk the trail by himself? Without or without walking beside him. Yes, you can. You can make your own choice in this; however, I do use a rope for the following reasons.

CONNECTION, ROLE REVERSAL, AND RECIPROCITY

When you are holding the rope without putting tension on it (i.e., with a 'smile' in the rope), the connection through the rope helps you as a handler to feel connected to the horse and the tracking. You are trying to see if the horse can solve the tracking puzzle with you as a team. What I also like about this is that the roles are reversed. The horse is doing something the person cannot do: He is taking the lead and the person is following. This is very uncommon in the world of horses in which training and control of the horse are usually central. The person is usually in the lead, giving directions, and the horse follows. That is why it can sometimes be a little uncomfortable for handlers to be the one who follows, trusting the horse and his abilities. As such, you sometimes see handlers carrying more tension in their bodies, checking anxiously to see if the horse is doing it properly, being ready to jump in and take over, trying to steer the process by the rope and their body language, becoming irritated when the horse is not as fast as they would like, or being surprised to see their horse doing well. This is all well intentioned, of course. You are far less exposed to these experiences and teachable moments if you do not use a lead rope.

This is also true for the horse. Horses feel this connection too, and to them it is an entirely different, enriching experience to be allowed to take the lead in this situation. This enhances the reciprocity in the relationship. I use it as an instrument to improve or establish the relationship between horse and handler, not based on any philosophy or theory, but on real-life experience. For horses who have bad experiences hand walking with people by a rope, this can also be a useful tool to turn those associations towards the positive.

IN PREPARATION FOR TRACKING OUTSIDE THE COMPOUND

In several countries, if you want to take your horse tracking outside the area of the stables, you have to be mindful of laws requiring you to keep your horse on a rope. In order to practice using a rope in the woods, you start using a rope in the arena from the beginning, especially because tracking has a different dynamic than horse and handler are used to.

STABLE RULES

In some stables, you are required to keep your horse on a rope in several places inside the stable area, such as in the arena.

WHEN DO YOU LET YOUR HORSE OFF THE ROPE WHEN TRACKING?

It could be the case that your horse associates the combination of the rope, the halter, and a person walking beside him with walking exercises or obedience. It could be that he has an injury, or that it is a good idea to take off the halter or forego the rope for some other reason. This depends on the individual case.

12.5 A HALTER

While tracking, the horse inhales large quantities of air. As he does this, the nostrils open wider, but the tissue around the nostrils and mouth moves as well. It contracts and expands (**Figs. 12.8–12.9**, **Figs. 12.10–12.11**). I want to impose as few barriers as possible to this, which is why I track using a halter or a bitless bridle (not the bitless bridles in which a hinge mechanism presses down on the nose, like a Hackamore). However, the horse can also track with a bit in his mouth. Just be careful that the nose strap is not too tight. You should be able to easily fit two fingers in between the strap and the nasal bone.

Fig. 12.8 The tissue around the nostrils contracts.

Fig. 12.9 The tissue around the nostrils expands.

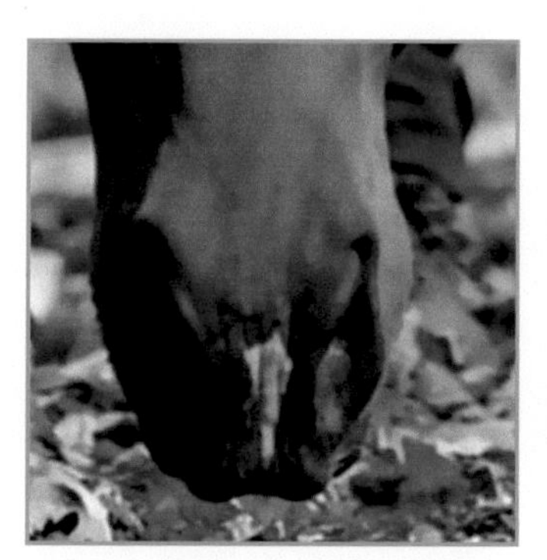

Fig. 12.10 The tissue around the nostrils contracts.

Fig. 12.11 The tissue around the nostrils expands.

12.6 WATER

Tracking is thirsty work! Always have water at the ready, allowing the horse to drink before and after tracking.

REFERENCE

1. University of Exeter (2018) *Research into equine vision leads to trial of new fence and hurdle design to further improve safety in jump racing*. https://www.exeter.ac.uk/news/featurednews/title_686716_en.html

13 *Laying a track*

13.1 TIPS FOR BEGINNERS

Decide in which area you want to track, likely an area in which your horse feels at ease. Find an area where no one, including you, has walked yet: a 'clean' canvas. If you are laying multiple trails, you keep moving your smeller and your bag to a new clean area. This will sometimes require you to plan ahead with regard to how to use the space and walk in and out, so that you do not 'contaminate' future clean areas for that session.

Determine the direction of the wind. In the beginning, try to lay the trail in such a way that the wind does not blow the scent of the food inside the scent bag toward the horse. After all, you want to teach him to follow the footstep trail, not to air scent.

See if there are scents that might distract your horse, and decide if you will or will not use them. For instance, if there is a pile of droppings in a corner of the arena, it is better to use another area of the arena first so your horse does not get distracted by it.

If you want to work with arcs or turns, make sure they are not too sharp. Sharp turns that are close together can cause your horse to jump from one scent trail to another. You want to avoid this in the beginning.

Keep a diary, so you can look up which trails you laid the last five times. Make sure you do not hide the bag in places where you have hidden it before. The horse's cognitive map is so good that, after a couple of sessions that follow a predictable pattern, it is quicker for him to find the bag by going around checking the familiar hiding places than by tracking.

Here is an example of how you can lay trails **(Figs. 13.1–13.17)**. In the top picture, you can see where the trails are at the location (in order to have a clean piece of ground for every trail). In the picture at the bottom, you see an enlargement of this trail. The solid blue line indicates how you can enter the space, and the blue dotted line shows how you can exit it. The dotted orange line is the scent trail. When the trails start to get longer, you will no longer be able to lay as many trails in the same space. Remember that this is a rough outline just to give you an idea. The point is not for you to lay this exact trail, and all trails certainly do not have to be completed in a single try. See what your horse is capable of (see section 14.7). In addition, the location, the wind, you, your horse, and how he follows the trails determine how you continue. Use the tips from this book to make these calls. In these examples, assume that there is no wind.

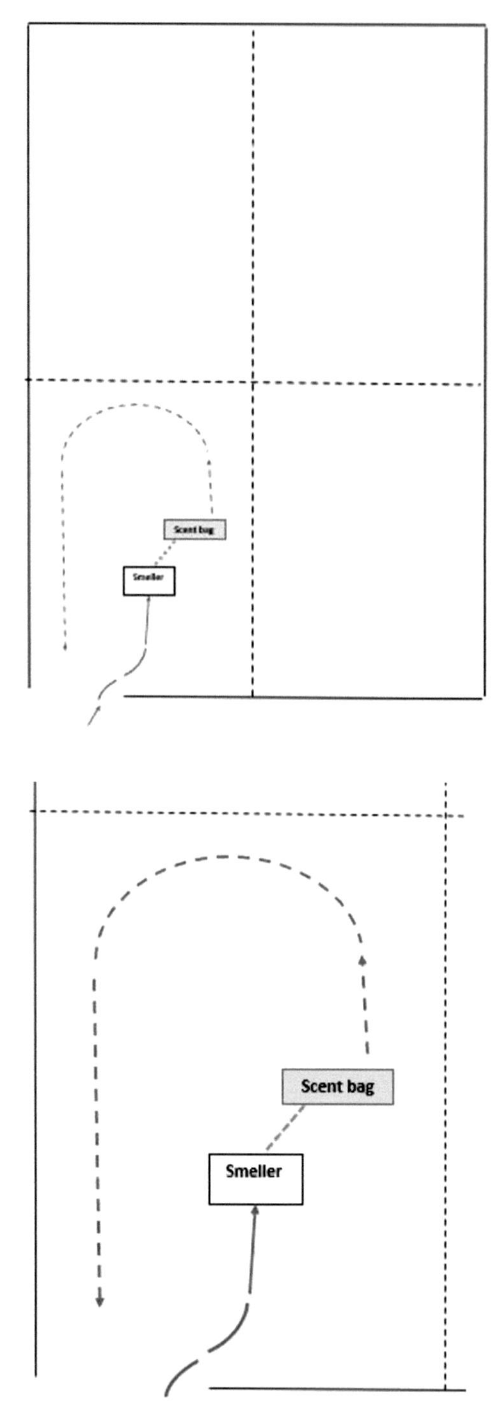

Figs. 13.1–13.2 A first trail could look like this. The scent bag is placed around 30 centimetres away from the smeller.

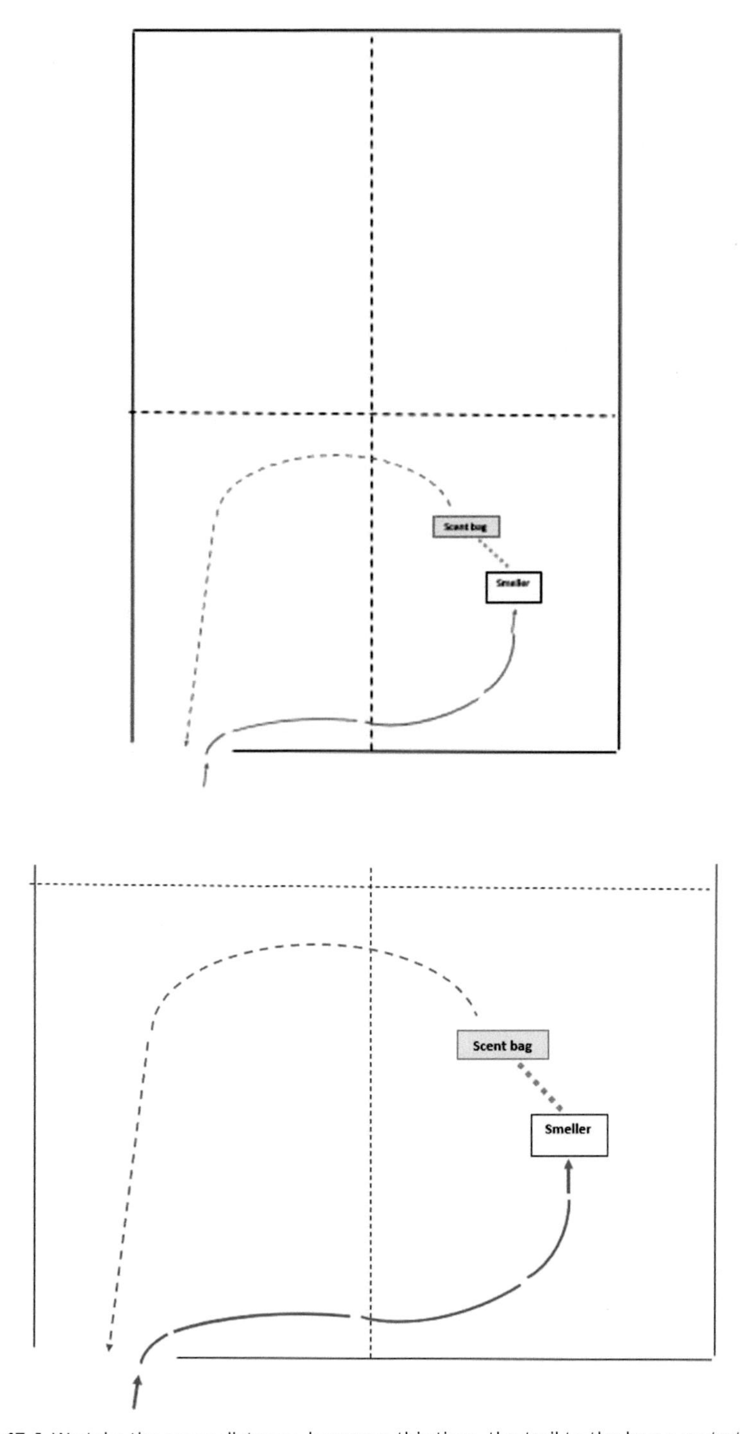

Figs. 13.3–13.4 We take the same distance; however, this time, the trail to the bag runs to the left instead of the right.

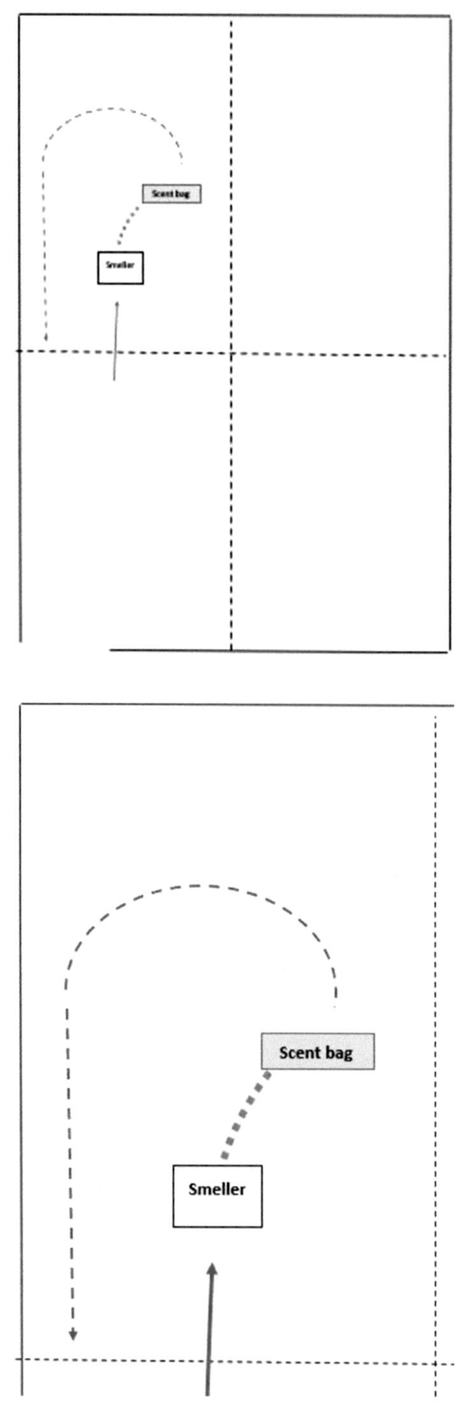

Figs. 13.5–13.6 If everything is going well, the next step can be to lay a trail that is a bit longer, perhaps one of around 50 to 60 centimetres in length with a slight curve to the right.

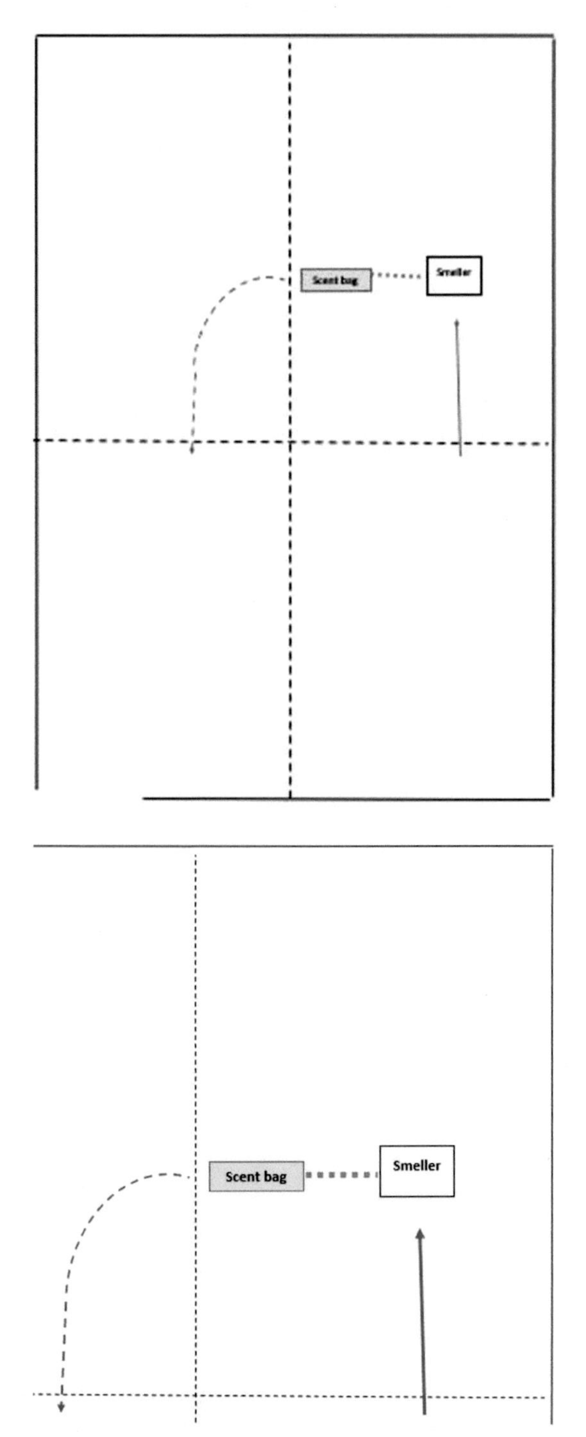

Figs. 13.7–13.8 The next trail is 30 centimetres again; however, the trail starts to the left of the smeller to prevent the horse from automatically walking straight ahead when he sees the smeller.

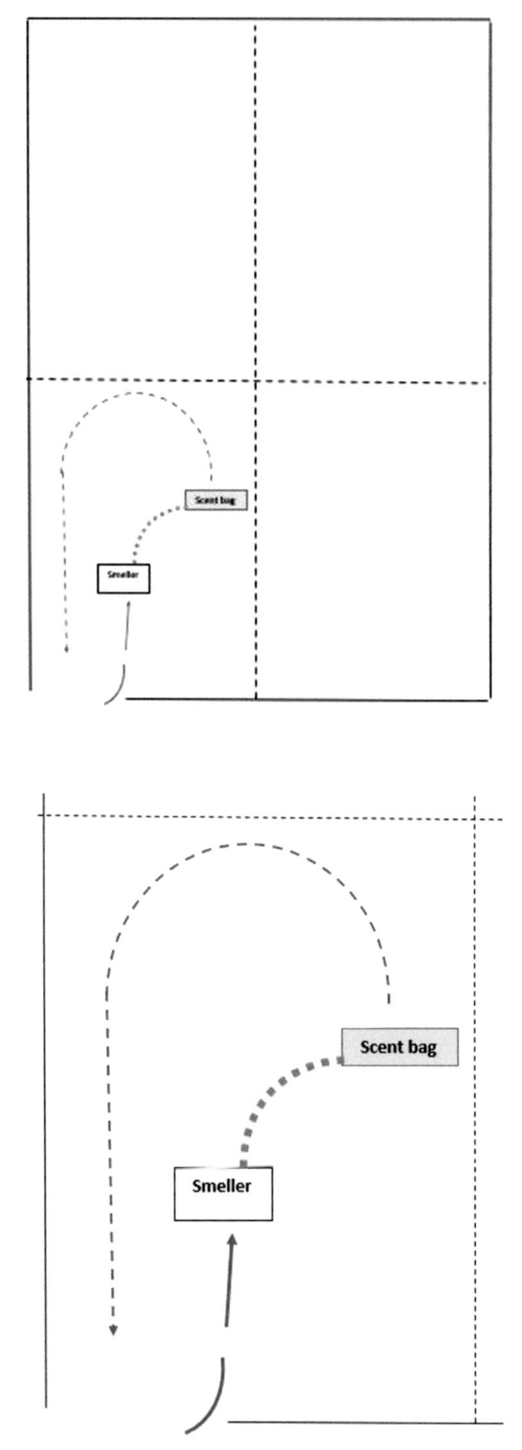

Figs. 13.9–13.10 If your horse is ready for a bit more, you can make the trail a little longer – around 90 centimetres.

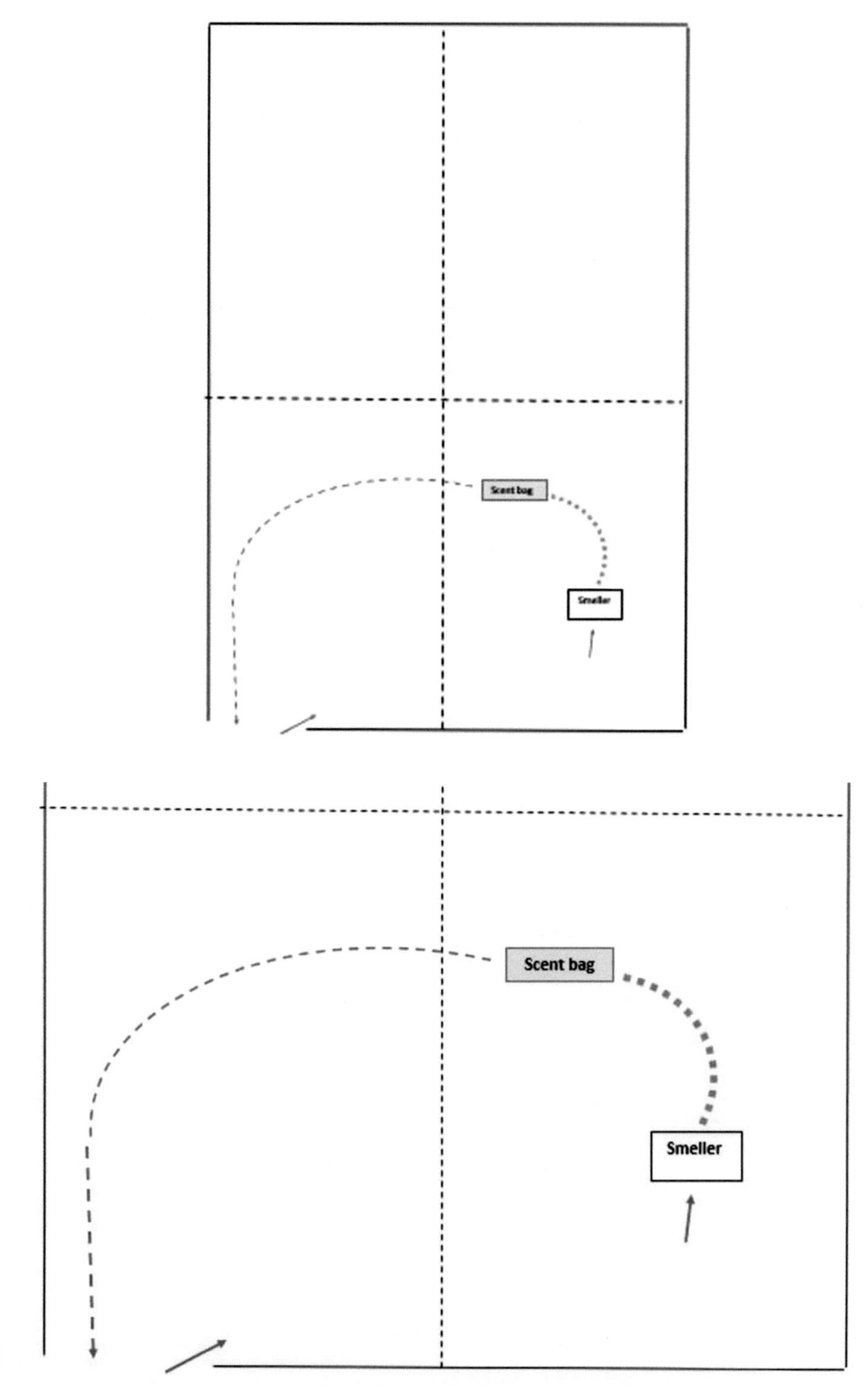

Figs. 13.11–13.12 This trail is a bit longer still, and this time with a left turn.

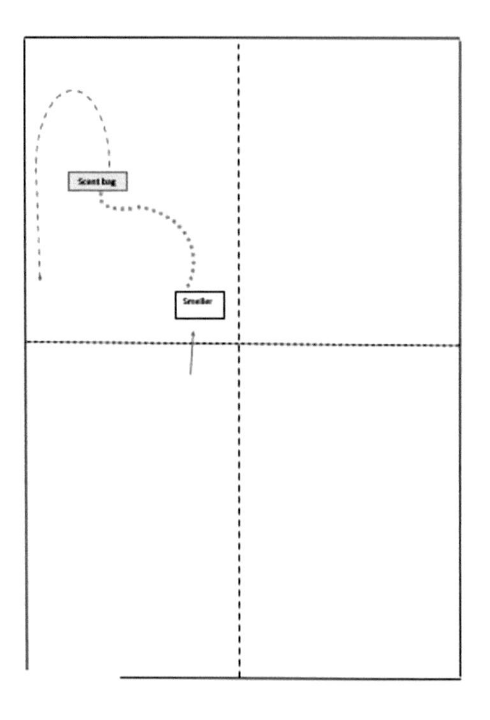

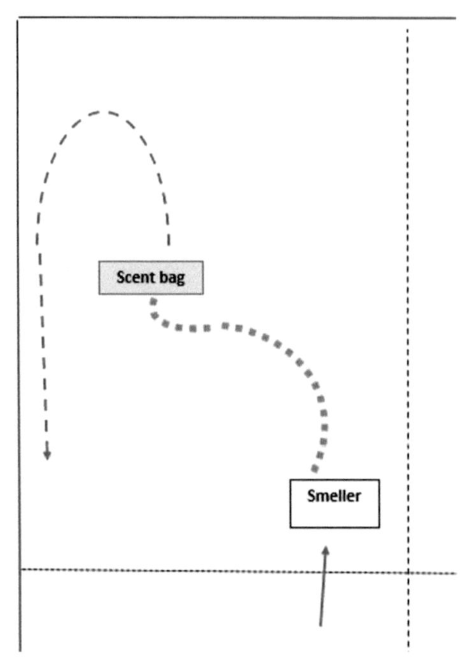

Figs. 13.13–13.14 Depending on your horse, you can make the trail even longer here or create a trail with a different curve. Here is an example of a trail that is lengthened a little again.

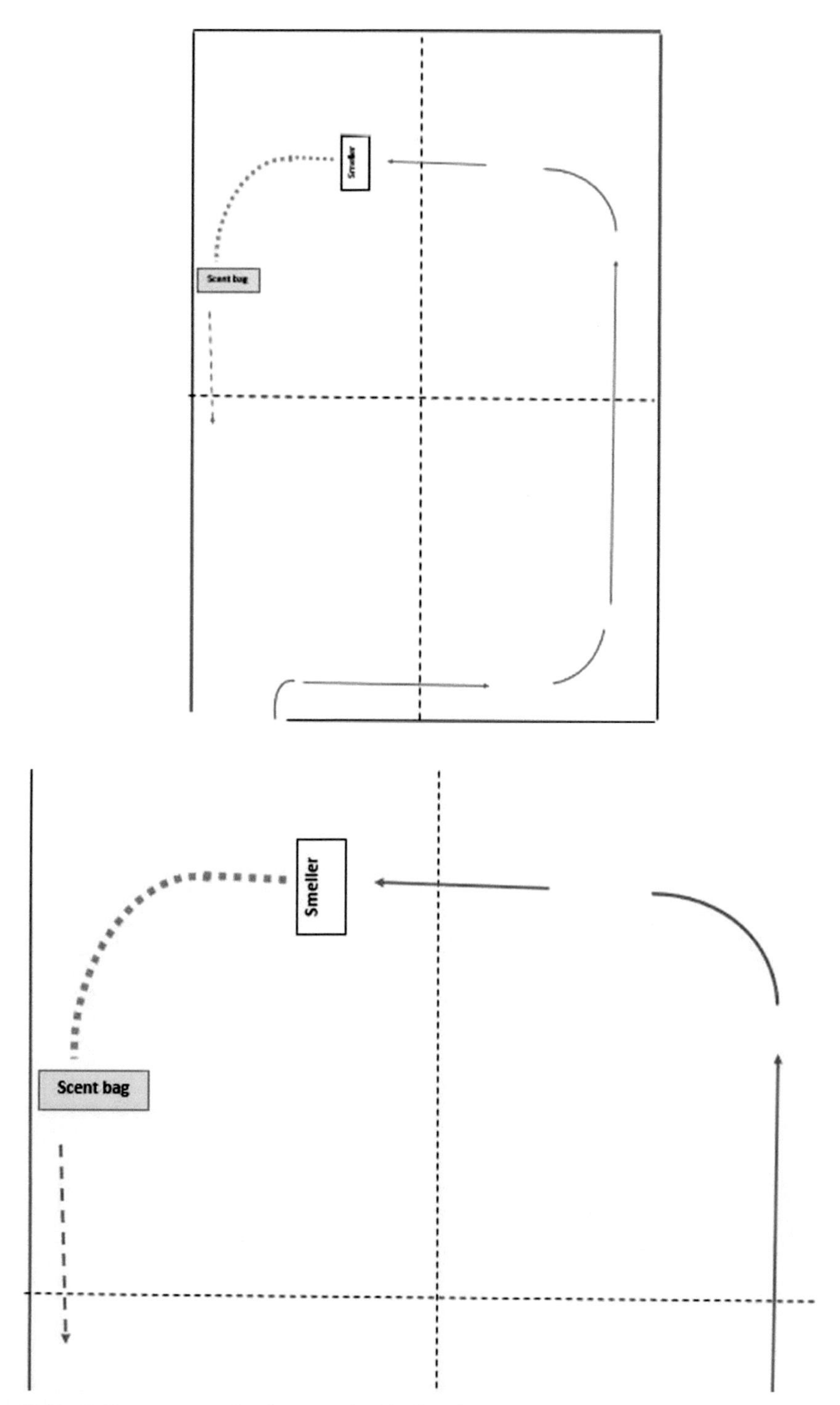

Figs. 13.15–13.16 Let us see what happens in this situation.

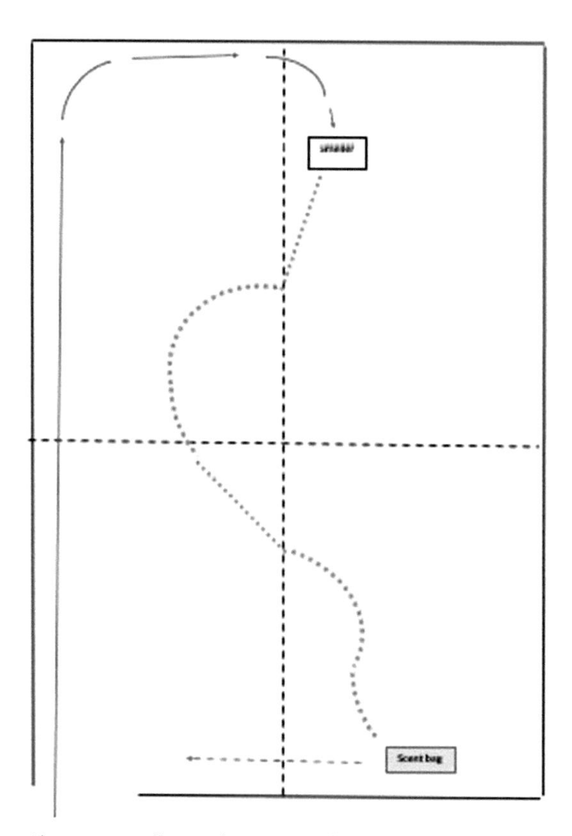

Fig. 13.17 You can lengthen your trails step by step until, perhaps, you are laying this trail.

Of course, you can relax these rules as your horse becomes a better tracker. You might even want to start looking for challenges, such as laying a trail between piles of droppings that your horse is allowed to smell before continuing with the trail, or choosing an area to track in that has lots of distractions. In any case, always look to your horse. His abilities and enjoyment determine what you will do.

13.2 IS YOUR HORSE SOCIALISED TO THE SCENT BAG?

Most domesticated horses eat from all sorts of troughs and buckets. They are used to stimuli around them and have mostly positive experiences in their interactions with humans. So, for most horses, it is no problem to eat from a scent bag that is new to them, especially because it is a comfortable size and not restrictive.

It is possible, however, that you have a horse who really has to get used to every new stimulus. This could be a scent, a change in routine, or a blanket that is suddenly lying in a different place. Such a horse would view the scent bag as a whole new object, different from the bucket or trough he is used to, which can result in ambivalent behaviour. For instance, the horse could show a mild or strong flight response when finding the scent bag. It would be a shame if, on your first track, your horse smells the food and locates the bag, making a really good run, but when you go to retrieve the bag, he is spooked by your movement, by the bag itself, or is afraid to eat from the bag, thus ending the positive tracking in a negative result. You want to prevent this, of course.

If you are not familiar with a particular horse and start tracking with him for the first time, or if you are not sure how your horse will respond to the first tracking attempt, err on the side of caution, and start with these exercises *before* you start your first trail.

STEP 1

Before you start laying a trail, check to see whether your horse is afraid to eat from the bag with and without wearing a halter.

STEP 2

Check the horse's motivation. If, while holding the filled scent bag, you take a couple of steps back from the horse while he is eating out of the scent bag, does he walk toward the scent back in one, two, three paces? (Note: Do this only once. The horse should not get the impression that food is being taken away from him.)

STEP 3

Also, before you start laying your first trail, check to see if your horse is okay with finding a bag full of food on the floor. You can do this by putting the bag on the floor, bending down, and picking it back up. If your horse is not completely relaxed, as an intermediate step, you can do the following:

- Show your horse the filled scent bag and let him smell it.
- Put the bag down on the floor right away, and let your horse smell it or eat from it.
- Calmly move toward the bag and pick it up while the horse is eating.
- Let the horse eat.

STEP 4

Have someone else place a filled scent bag inside the arena and see if your horse walks toward it on his own. If your horse sniffs the scent bag, you are able to pick it up and open it, and the horse starts to eat from it without tension, this is a good indication that you can proceed.

Important: Looking at your horse and what he needs is more important than sticking to the steps. If your horse needs repetition and intermediate steps, do this. If you can skip steps, do so

How Kosmonaut's first steps went

(**Figs. 13.18–13.20**, **Fig. 13.21**, **Figs. 13.22–13.25**, **Figs. 13.26–13.28**).

Figs. 13.18–13.20 Top picture: Laura offers Kosmonaut food in the scent bag. He eats from the bag. In terms of posture, you can see that his body is still somewhat elongated, with his right hind leg set back a little. This is a sign that he could be more relaxed. Middle picture: Laura walks backward a few steps with the filled bag, and Kosmonaut follows. Bottom picture: Kosmonaut eats from the bag, every once in a while lifting his head to chew and look at the bushes.

Fig. 13.21 We also see if Kosmonaut will eat from the bag while wearing a halter. In the picture above, he has just walked up and is putting his nose into the bag.

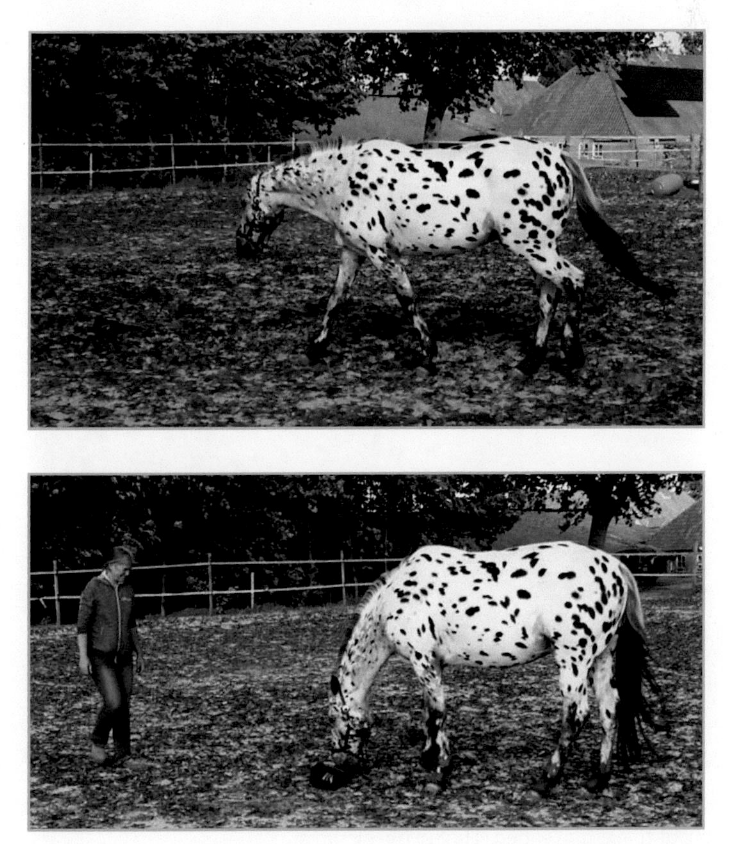

Figs. 13.22–13.25 Laura places the filled scent bag in the middle of the paddock. In the series above, Kosmonaut can be seen walking toward it. He sniffs it.

Figs. 13.22–13.25 (Continued) Laura approaches, bends down, and picks up the bag. She opens it, allowing Kosmonaut to eat (the eating is not in the pictures).

Figs. 13.26–13.28 To reconfirm Kosmonaut and check again whether he is comfortable enough, we place the bag in the paddock again so that he can walk toward it. Laura can pick it up, and he can eat from it.

Important: When another person is putting down the bag in this last step, do not restrain the horse when he approaches the bag before the other person is done. You do not want the horse to think that he cannot take initiative or that he cannot approach the bag. Quite the opposite. He needs to take initiative.

Important: You have your varied reward that your horse considers valuable in the bag when you are working to socialise your horse to the scent bag. At this time, you can fill it with half the amount you would when you start to track, but do not be stingy. The horse can spend at least a minute eating it and licking the inside of the empty bag.

13.3 IS YOUR HORSE SOCIALISED TO THE SMELLER?

It is rare for the domesticated horses I track with not to be socialised to a piece of cloth on the ground. So, normally, we do not check this. Without preparation, when a horse sees a piece of cloth on the ground, he will approach it without difficulty. However, if you have a horse that has seen very little in his life, start doing enriched environments with towels and pieces of cloth on the ground in preparation so that he can get used to pieces of cloth on the ground and will want to investigate the next time he sees one.

14 *Tracking*

Step 1: The beginning

It is important that you do not let your horse see you laying the trail. In the beginning, this will not be a problem, but after a few times, your horse will walk straight to the hidden scent bag. How to get started (**Figs. 14.1–14.12**).

Fig. 14.1 Collect different kinds of foods your horse really likes, and put them in the scent bag. Find a good place to hide the bag.

Fig. 14.2 Lay down the smeller.

Figs. 14.3–14.5 Step on the cloth firmly a number of times wearing shoes you wear often and have textured soles.

Figs. 14.6–14.7 Walk to the location where you are going to bury the scent bag, setting your feet firmly on the ground with every step.

Figs. 14.8–14.11 Bury the scent bag by digging a small hole with your feet, hands, or a small shovel or by laying the bag flat and covering it up with sand or debris found on the ground.

Figs. 14.8–14.11 (Continued) Cover the scent bag up with sand or debris found on the ground. Very important though: **leave the mesh uncovered and open to the air!**

Fig. 14.12 Leave the track with big steps. You are ready to get your horse.

14.1 WHAT DO YOU NEED TO MAKE THE FIRST TRY A SUCCESS?

14.2 THE HORSE IS SOCIALISED TO THE SCENT BAG

The bag can be picked up out of the ground. You have done exercises for this from section 13.2.

14.3 THE HORSE IS USED TO THE HANDLER AND THE POSITIONS HE TAKES WHILE FOLLOWING

When you are tracking together, it is necessary for the horse to be accustomed to the handler. If the horse knows how this person smells, moves, talks, laughs, coughs, sneezes, and what he or she looks like, he will not be startled and can focus on the tracking. The horse is also used to the handler walking at his flank in various positions, or directly behind him. The handler must be able to walk on both the left and the right side of the horse. In the arena or the pasture, you will see that you often walk at your horse's flank. However, if the horse makes a fast turn on his axis, it can happen that you are standing behind him for a moment, before returning to his flank. If you also intend to go tracking in the woods, which often involves walking on narrow paths or through bushes, then you will want to be able to walk behind your horse.

EXERCISES

If you want to familiarise your horse with the different positions you can take while walking together, there are several ways to do this. Here are a few that I use.

1. You can stand next to your horse and walk alongside him as he is grazing at pasture. As you are doing this, you can walk with him, calmly switching positions every now and then. To many, this may seem like a very passive way to 'do something' with your horse, but it is definitely useful. It also increases your horse's bonding behaviour toward you. After all, eating together in a group is a bonding experience. That you are not eating grass yourself does not diminish this effect.
2. If you and your horse are ready for it, and it can be done safely, you can take your horse for a walk, again assuming various positions by his side. You can do this while walking or while he is standing still to graze.
3. You could try not tying your horse up while grooming him, but instead put him in a paddock. Begin to groom your horse while standing close to him, and pay close attention to his body language to see what he likes and what he does not like. You can sometimes lean into him a little as you give him a good scratch in his favourite places (to mimic how horses groom each other). You can build toward this at the horse's hindquarters. If both you and the horse are not comfortable with it, move toward it in small increments. Hold your hand and arm out behind his hindquarters and then remove it again, or put your leg behind his hind leg. Build all this up slowly until you are able to calmly walk or stand behind him. The advantage of doing this in a paddock is that your horse can step aside or walk away if he does not want to do something anymore. It is good that your horse indicates this. I see stepping away as a valid reaction, and I temporarily stop what I was doing that made the horse step away, allowing the horse to recover and get back into his comfort zone. I then see if I can break up what I was doing into smaller exercises and start to practice again. I do not turn this into an obedience exercise, nor do I reward the horse with bits of food. I am just focused on seeing the horse's reaction in the form of body features and signals (I do not want the emotions he feels in response to getting bits of food to muddy the waters). I proceed in very small steps. I make sure that I know where my horse likes to be petted, and I use this to give him little reward breaks.
4. You can do the haystack exercise, which is in the appendix.

Important: I absolutely do not use a whip or flick a rope to drive the horse on before me. I do not want to cause tension or trigger a flight mechanism in any way, no matter how mild. In order to track well, the horse needs a relaxed body and mind.

With regard to the exercises described above, I base my advice on the majority of horses and horse owners I visit. These horses are all accustomed to a particular person and walking with that person. If problems do exist, they are of the sort that, for instance, the horse has learned that he has to stay behind his owner, or needs to learn to relax more when his owner is standing or walking behind him. It could be that the horse stops and refuses to move or walks too fast, pulling his rope tight. These are challenges for which the suggestions above are useful.

If, however, you have a horse who is entirely unapproachable, then it would be good to do a consultation with a behaviour specialist. That way, you can arrive at a whole array of measures, of which exploration and tracking can be a part.

For such a horse, I would recommend a treat search (see Chapter 18 Treat search), moving along with the horse at a distance, and standing or walking parallel to him (this is also a good exercise if the horse displays aggressive behaviour at feeding time). You do this at a distance he is comfortable with and slowly come closer over time.

14.4 THE HORSE IS SOCIALISED TO THE LOCATION

For a successful first tracking run, it is important that the horse feels safe and relaxed enough in the area in which he will be tracking, has no separation anxiety, and is confident enough to focus on the tracking job in front of him – a job he is still learning to perform at this point.

A relaxed horse has no trouble lowering his head and neck and keeping them lowered because the relaxation of muscles and tissues enables him to do this. As we have seen in Chapter 7 about biomechanics, a horse whose tension level is higher has a mid-high to high neck position and short, hard muscles. He is restless in his motions and is mostly scanning his environment from this higher head-neck position. If he even moves to smell the smeller, it will be brief, and he will not be able or it is much harder for him to make the transition to sustained scenting and tracking the scent with a low head position.

This is why it is important to pick your scenting location with care. You can walk around with your horse and read his body language to see where he is more and less comfortable. You can also choose to give your horse the opportunity to get to know the tracking location. You could take him for a walk there, let him eat hay there, or let him graze if possible. After all, prolonged chewing has a relaxing effect. Perhaps you could take him there with a buddy so that they can both explore. Or you can have a buddy present when you do the tracking. You can also do the exploration games described in Part 1.

If your horse feels at home only in his own pasture, start doing tracking exercises there, do a treat search, or play some of the exploration games described in the first part of this book. As times goes on, you can gradually move the location of these games.

CAN HORSES TRACK WHILE FEELING TENSION?

It is not impossible to track with horses who are tense, but it all depends on the level of tension and how good a tracker your horse is. Tracking is a calming activity. It is also an important way to enable your horse to calm himself down when he feels tension and focus on the trail, but learning this takes practice. And if he is relaxed or only mildly tense, a novice tracker stands a much better chance of learning all this.

A QUIET DAY: NO NOISES OR ACTIVITIES THAT CAN INTERFERE

In light of what we just discussed, it is obvious that you should choose a quiet time for your first tracking sessions. The horse has to be able to concentrate, and it would be a shame if he were startled while tracking, if someone were to yell out, if a container were being emptied, or if his buddy were to walk by, prompting him to want to follow. Also make sure your phone is set to silent mode so that it does not make any noise.

If the horse is startled or distracted, he will no doubt get over it, but sometimes that takes a while, and you can avoid it with good planning. Then once your horse has the hang of tracking, these noises from the environment will generally no longer be a problem, and he will not be distracted.

14.5 HANDLER SKILLS

POSITIONING AND HOW TO HOLD THE ROPE

Fig. 14.13

As a handler, you walk by the horse's flanks or hindquarters. That way, the horse can smell optimally and is not distracted by your scent directly beside him. Also, he can easily turn and decide his own path this way. And we are literally the followers, so it is only right that we take that place.

We keep the rope at a length at which it does not touch the ground, including the part with the hook. Sometimes you will have to pay extra close attention to do this, especially when walking curves or when your horse lowers his nose so much that the rope almost inevitably touches the ground. You want to prevent the horse from stepping on it. Try to hold the rope in such a way that it has a light arc to it (i.e., a 'smile' in the rope). You do not want to put pressure or tension on it. If you find this difficult, and you want to make sure that you are not obstructing your horse during the first tracking sessions, you can also pull the rope through the halter's chin strap. Of course, you can only do this if your horse does not have a problem with it and walks along and stops easily (**Figs. 14.14–14.15**).

Fig. 14.14

Fig. 14.15

WALK SLOWLY: TRACKING SHOULD NOT RESEMBLE A WALKING EXERCISE

It seems to be a habit for handlers that when they are walking with a horse, they move fast. Of course, this could be because the horse walks at a faster pace, but I see very few people strolling along, taking seemingly unplanned turns here and there. It could be that this behaviour stems from the way we were originally taught to handle horses: the person always had to take the lead, and he or she was not supposed to grant the horse any freedom or succumb to any of his desires. Modern research shows that linear dominance does not occur within existing semiwild herds, and fortunately, more and more handlers are looking for a balance in which the horse also gets a say and does not merely have to follow and please his handler.

What does this mean for tracking? Consciously walk a bit slower than you normally would when you enter the tracking area and approach the smeller. The slower pace immediately allows the horse to take in his surroundings more consciously and see the smeller. The slower pace also ensures that the horse does not fall into an automatic familiar walking pattern in which he simply follows and does not respond to his environment, which, in this case, would include not noticing the smeller or developing initiative.

If your horses pass by the smeller, try the approach again, and stop your horse in front of the smeller. When you do this, make sure the smeller is not directly in front of the horse's forelegs. Have him stop roughly a metre in front of the cloth. That way, he can see it properly and act accordingly.

BE CAREFUL NOT TO GIVE INDICATIONS THROUGH VOICE OR BODY LANGUAGE

During my lectures, we always do a number of exercises. One of them always seems very difficult to the participants at first, but when it is over, they are all surprised at how easy it was. Two people do the exercise together. One of them, the 'horse', covers his eyes so he cannot see while the other person lays down three or four objects in an area of 3 to 4 metres. In his head, this person chooses one object that is 'special'. The 'horse' then has to guess which object this is. There are a few rules for the person who placed the objects: he cannot point with his hands, arms, or shoulders, or turn his body. He also cannot talk or use his breath to give clues, but he *can* use facial expressions. Every human horse guesses the special object easily. By themselves, our facial expressions can be very informative and send a lot of signals.

Just like a number of other exercises we do during lectures and workshops, this one makes the participant aware of how much body language he uses or can use during tracking. And this is important. Tracking should not become a game in which the horse reads the handler's body language instead of the scent trail to find the bag. So be careful not to give the following clues when you are following your horse when he is tracking:

- Looking at the location of the bag.
- Turning your body in the right direction.
- Leaning your body forward when you are both standing still, thereby indicating a direction.
- Pointing with your hand, arm, or foot.
- Walking faster and more enthusiastically when your horse is going in the right direction.
- Walking more slowly and sluggishly when he is going in the wrong direction.
- Sighing or holding your breath when the horse is going in the wrong direction or passes by the bag.
- Giving indications through the rope, even unconscious tugs.

- Making a happy, sad, or angry face when he is going in the right or wrong direction.
- Pulling up your shoulders when the horse is going the wrong way.
- Standing still when the horse is nearing the bag: He will learn when that happens that he is close to the bag, as well as the rope-length at which he will find it.

What should you do?

- Try to be as relaxed as possible when walking, with gentle steps and musculature.
- Keep your shoulders low and relaxed.
- Try to stay balanced and not lean forward.
- Relax your face.
- Keep your breathing as even as possible.
- Move along with the horse as he turns his head and body and follow him.

USING YOUR VOICE

A quick note on how to use your voice. The horse has to concentrate very hard, especially in the beginning, he will be using a sense that he normally does not consciously use. He needs to have a lower head-neck position. If your voice is high, and out of enthusiasm, you shout 'Yay!' or 'Well done!', this can have an adverse effect, causing the horse to lift his head. When I am tracking with a horse, I am silent. If I do use my voice, I adopt a lower, softer tone. All of this can take some getting used to, especially when you are tracking and you know where the bag is hidden.

So, can you *never* help your horse? Preferably not, but there are exceptions. For instance, if your horse is walking so fast that it is uncomfortable for you to follow him, which might cause you to fall if you were walking through the forest. Then you can gently slow him down, not so much that he stops tracking, but enough to slow his walk a little. If at all possible, however, you should let the horse track on his own, so that he is not dependent on people.

THE HANDLER DOES NOT HELP OR INTERFERE

When you start tracking with your horse, you assume two roles: director/planner and follower. You plan the tracking session like a director, taking all the day's variables into account, including what your horse is like, what location to use, the direction of the wind, what trails you have laid before, what reward to give in accordance with the difficulty of the trail, and so on. In this moment, you are the planner. However, as soon as you start the tracking session with your horse, and he takes the lead, you are the follower. You see how your plan shakes out and what your horse does. Then, depending how it went, you make a plan for the next step.

Important: If, in your eyes, the horse makes a mistake, for instance by turning left instead of right, do not stop him. It is important for the horse to develop a tracking technique of his own. Some horses change direction to make sure the trail really is not there, after which they return to the 'real' trail. Or, if they have found the bag, they do another small round to make sure. I often see that when the handler stops the horse or gives him clues, it takes the horse out of his own thinking and tracking process, after which he will stop searching altogether.

Photo series: The first trail of the tracking lesson had been laid. Aysa was eager and wanted to find the bag quickly, causing her to walk too fast and miss the left turn. (Note: The trail was laid more deeply on purpose here so that I could show it to the participants.) (**Figs. 14.16**, **14.17**, **14.18**, and **14.19**)

Fig. 14.16 Aysa took large strides forward, causing her to miss the turn to the left.

Fig. 14.17 Upper right picture: She realized that she has lost the trail.

Fig. 14.18 She started to circle and air scent to find the bag.

Fig. 14.19 She found the bag.

The second trail we laid during that class was in a different place, but it had the same kind of turn to the left, because we wanted to see if Aysa adapted her tracking strategy (**Figs. 14.20–14.23**).

Figs. 14.20–14.23 Aysa started to track again. This time, she took very small steps, keeping her pace much slower than before. Her nose was lowered far down to the ground. She did not want to miss a speck of the trail, especially not the left turn; and she did not.

Figs. 14.20–14.23 (Continued) She went straight for the bag.

The example above demonstrates that tracking can be challenging for the tracker and the handler. It is a process of trial and error. You can be surprised by what your horse does or does not do. And of course there are different ways to define a successful trail, even if the horse does not tracks in exactly the way you envisioned. If the horse does not find the bag, but this teaches him that he has to move more slowly, or perhaps follow the direction of the wind more, it is still a successful trail. In this sense, a trail that does not lead to a found bag can still be valuable. What went wrong? What did the horse show? Is there something you have overlooked? The checklist in the appendix can help with this.

DO NOT CONTROL THE HORSE BEFORE OR DURING THE EXERCISE

Your horse should literally take the lead while tracking and walk in front. When I start tracking, and the horse is excited that the session is about to begin, I will hold him back as little as possible. I will walk with him a little faster or, if it can be done safely, even run along if he trots. I enjoy his enthusiasm. If you do obedience exercises right before or during tracking, such as the horse walking on your left or right, halting on command, or walking behind you, you can put your horse in an 'obedience frame of mind', which influences the tracking, in which he *is* allowed to make his own decisions. This can really interfere with the tracking, especially in the beginning.

14.6 FIRST CHALLENGES YOU MAY ENCOUNTER

Don't give up when you and your horse are just learning to track. My horse Aysa took three sessions to do anything resembling tracking. I had nearly written it off as nonsense, boring, not for Aysa and me. Luckily, thanks to Rachaël, we stuck with it. Once Aysa discovered what she was supposed to do and found out how good the reward was at the end of the trail, it became fun! Now, Aysa runs towards the smeller and starts tracking right away. Having fun together makes me so happy!

Martine Liefstingh

You have read this book and taken its tips to make tracking a success to heart. Your horse is socialised to the space you have chosen. You have your scent bag ready. You have tested your horse's top 10 rewards. You have filled the bag with variations of your horse's top 5 favourite foods. You have tested your horse, and he is not afraid of a bag full of food on the ground or of the smeller. You have studied how to lay a trail. The weather is perfect. You have prepared everything. You get your horse and lead him to the smeller, bursting with anticipation, and your horse smells the smeller, raises his head back up, and walks away to the other side of the arena, looking for something to eat. In other words: total failure! What went wrong?

First of all, do not panic. This happens to some horses. The first trails are the most difficult for a horse. After all, he is expected to do something he has never done with you before. This is easier for some horses than others. It also depends on the horse, of course: how he is feeling that day, what his life has taught him, what his learning experiences have been like, and the emotions he has learned to feel when it comes to the stimuli to which he is exposed.

And, as said before, the majority of domesticated horses have never consciously used their noses to find food. As such, their noses are not yet turned on for this task. Perhaps, added to this is a lack of experience in discovering objects with their senses. They might not yet be curious enough to do this. Building suitable enriched environments, in the period prior to the scenting, could be of good help for these horses.

Below are a few possible reasons why your horse is not following the trail from the smeller to the scent bag on the first try in particular.

- The horse has never used his/her nose before and has to find out how to use it again.
- The horse does not take initiative. This is most common in horses that are described as especially docile. They have learned to submit to our wishes and lives and are obedient. Not taking initiative is a part of that. As a coping mechanism, these horses can also become passive when they encounter new

stimuli. The 'not knowing what to do' turns into standing still and no longer taking any initiative.

This tendency to no longer take initiative can also be tied to a place and time. Your horse could associate the arena or the paddock with training. The above is also true of horses who have been taught not to pass by their handler and always walk behind him/her. I have also seen that these horses do not make turns in the direction of the handler (a turn that some people view as a dominance turn into the handler's personal space, thus punishing the horse for it). These horses have to overcome an additional hurdle to be able to stand or walk in front of a handler or make such a turn if they want.

- The horse stands beside the owner. As a consequence of not knowing what to do, this horse stands next to his handler. The horse can find this quite pleasant (watching TV together) and he does not feel the need to take initiative.
- The horse has learned that exercises should be done in proximity to the owner. These horses seem reluctant to turn away from their owners if they think there is a joint activity in the offing. These horses often like to stand with their heads close to the handler or facing the handler. It can also be that they associate this handler with a food reward, causing them to keep their heads close to him or her.

Note: The first scent trail can also be more difficult for horses who have done a lot of games that purposefully contain physical muscle training, such as picking up objects, doing the Spanish walk, sticking their heads through an object, and so on. Horses who have been or are being trained using the technique of shaping can have trouble during their first tracking session. Shaping is a training method whereby the trainer has a final goal in mind and comes up with intermediate steps that will lead to this goal. The horse can then try out behaviour on his own, and the trainer will reward him if the behaviour corresponds with the desired trajectory and intermediate steps of the exercises and criteria. The idea behind this is that the rewards in the interim motivate the horse to stay involved and reach the final goal. When many of these horses get to the smeller, not knowing what to do, they seem to repeat the physical tricks they have learned such as the Spanish walk, picking up the cloth and shaking it, and bowing down.

These behaviours are far removed from using their noses, which is an entirely different discipline. In addition, the horses who are used to getting a lot of treats along the way can suffer from impatience. After making the smallest move, they are already looking at their owners for a reward. In that case, the first trail to the scent bag can be a much more intensive task than they are used to.

I believe it is impossible to reward the horse for the tiniest intermediate steps during these first tracking attempts. We do not know exactly when the horse starts using his nose and smells. When he does do this, it requires intense focus, which you do not want to break to give him a reward. When owners do give intermittent rewards during the first tracking session, I see that the horse associates this reward more with lowering his neck than with using his nose. As such, the intermittent rewards are counterproductive in this situation.

I discuss this more extensively because I meet a lot of people whose horses are well-trained in other specific behaviours who then expect the horse to be a good tracker as well. They do not expect this extra barrier. At the same time, horse owners are sometimes surprised that their somewhat 'difficult' horse is so good at tracking. The horse's wilfulness and 'disobedience' now serve him well.

- The horse does not take the food with the human present. Some horses who live and eat in groups are not assertive when it comes to claiming food but rather quietly eat along without anyone noticing. These horses can be hesitant to take the food when their owners are present. This could just be true of the first few tracking attempts; once they discover that the bag is really meant for them alone, they are very eager to find it (**Figs. 14.24–14.27**).

Figs. 14.24–14.27 Indy completed her third trail. Ristin was eager to see how she did and anxiously looked on to see if Indy would manage to find the scent bag. Indy stopped. She knew where the bag was, but she did not take it. It seemed like Ristin's posture was causing this hesitation in Indy. Maybe Indy though Ristin wanted the bag too, as if they were both waiting to grab it. (In the bottom picture, she also gave Ristin a weakened distance-increasing signal.)

Figs. 14.24–14.27 (Continued) This situation continued until the moment Ristin took a few steps back and turned away from Indy. Indy then pointed to the bag (not included in the series).

- The horse has a very short concentration span. This means he has trouble concentrating for long enough to find the bag. He lifts his head along the way and then has to start over. Socialisation to the environment and its stimuli is very important here. There is a fine line between not being able to concentrate and being distracted by the stimuli the horse sees and smells all around him. This can also cause a shortened tracking time.
- The horse is in pain. As we saw in the chapter on biomechanics, in order to track, a horse needs a body that is capable of it. If the horse is in pain, it can impede the tracking posture. This horse will lower his head and neck, but he will lift them back up fairly quickly. This might give you the idea that the horse has a short concentration span, but pain or stiffness could also be the reason.

As discussed in the chapter on biomechanics, I am not doing research that can link established ailments or diseases to tracking behaviour, painting a before and after picture with the use of x-rays.

The horses I track with do not have problems that have been clearly diagnosed by a vet, physical therapist, osteopath, or the like. In horses that were visibly stiff in the back and neck, had very sparse musculature, were badly coordinated, or were a little wobbly, I noticed that these horses had some trouble getting started, particularly on the first trails. They were absolutely willing, but they raised and lowered their heads more often. They sometimes walked away at a faster pace, only to return to the smeller. If I saw this stiffness, I had the horse repeat the first sessions of scent tracking a few more times to allow the horse's body to get accustomed to this new posture. In addition, some horses learned that they are perfectly able to track if they do not hold their noses close to the trail but above it, as low as the body would allow. As time went on, they were able to go lower and lower, or they kept using the higher tracking posture with good results.

If the first trail is a failure, see if these tips will help:

- Bury the filled scent bag downwind. That way, you are consciously choosing to let the wind help out a little after all
- Bury the scent bag 5 to 10 centimetres away from the smeller. Make sure it contains freshly cut items and foods with a strong scent.
- For this new attempt, replace the cloth. A new cloth means a new chance.
- Use a biothane rope that is as light as possible (you have already replaced your old halter with a soft, painless one).
- If you think your presence is keeping the horse from taking initiative, you can try to be as uninvolved as possible beside your horse. Turn your back to the horse and do not look at him, or if you do, only in very brief glances and with a soft expression, looking just to the side of the horse. You can also see if your horse displays different behaviour if you let someone else hold him, or you can let him do the first trail without a handler or off leash. The handler will then join the horse when he has found the bag.
- If you have met all the conditions and are still experiencing failure, then maybe what you need is patience, in a single session or across multiple sessions.

Picture series of Chess getting started (**Fig. 14.28**, **Fig. 14.29**, **Fig. 14.30**, **Fig. 14.31**, **Fig. 14.32**, **Fig. 14.33**, **Figs. 14.34–14.36**, **Figs. 14.37–14.38**).

Figs. 14.28–14.29 Chess saw the smeller and sniffed it.

Fig. 14.30 Chess stood comfortably next to Nanne. After a little while, Nanne led Chess back to the smeller.

Fig. 14.31 Chess walked across the smeller as if he were being lunged.

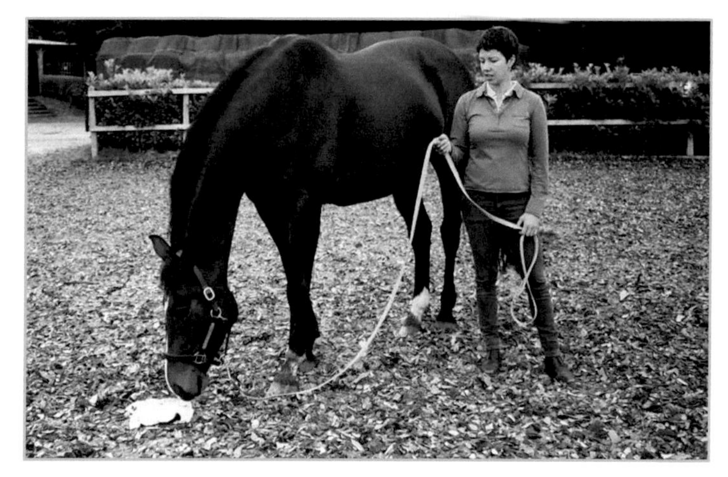

Fig. 14.32 Nanne stood with her back turned to Chess, and later stood with her body slightly turned away from him. We chatted a little.

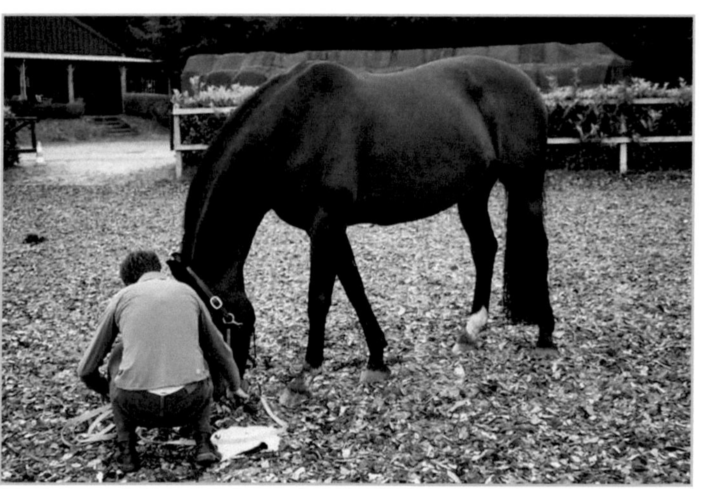

Fig. 14.33 Chess returned to investigating the cloth. She found the scent bag.

Figs. 14.34–14.36 During the next trail, we saw whether Chess started to track more easily when she was not held on a rope. This was the case, as she began tracking right away and found the scent bag, which was about 30 centimetres away from the smeller.

Figs. 14.37–14.38 Session 2, a week later. We saw how Chess did walking with Rachaël. The rope had been adapted and was super light, and the hook was very small. At first, Rachaël turned away from Chess. Chess found the bag, which was hidden 1.5 metres away from the smeller.

After this, Nanne, Chess's owner, can take over. Chess can now do it without any hesitation while Nanne is holding the rope. The reason to use a rope here is because Nanne also wants to track outside of the enclosed areas with her horses.

14.7 DIFFERENT WALKING PATTERNS WHEN THE HORSE IS TRACKING

There are different ways in which a horse can track. The horse may alternate these within a session (**Fig. 14.39**, **Figs. 14.40–14.42**, **Fig. 14.43**).

Fig. 14.39 With his nose low to the ground, he followed the trail without deviating from it.

Figs. 14.40–14.42 The horse followed the track, but he deviated from it by zigzagging from left to right with his head. It looked like he was checking both sides to see where he can and cannot detect scent.

Figs. 14.40–14.42 (Continued) The horse followed the track, but he deviated from it by zigzagging from left to right with his head. It looked like he was checking both sides to see where he can and cannot detect scent.

Fig. 14.43 Air scenting. The horse held his nose out a little. He can hold his head in any position, but usually this is between the mid-low and high positions. You also often see this pose when tracks are too easy for a horse.

14.8 THE HORSE HAS FOUND THE SCENT BAG

Hurray! The horse has found the bag. This is always a special moment, perhaps especially if you were no longer sure where you had buried it exactly. Most horses show that they have found the bag by rubbing their noses over it, sniffing it, or by scraping over it with a hoof in order to flip it out of the ground. I have only seen a small percentage of horses lift it up with their teeth. Some horses point to the bag by showing a seesaw lowering about a metre before the bag, which is more difficult to recognize. Others stand squarely with their heads and necks in the mid-high position, as if they are staring into the distance (not to be confused with a horse who is distracted because he really sees, hears, or smells something in the distance). Then there are horses who are very rushed and who are unsure, especially in the beginning, whether they are allowed to touch and take the bag. They can sometimes purposefully touch the bag, but because you, as the handler, do not give a verbal reward, they can keep walking. Generally, you do tend to get a second chance when the horse approaches and taps the bag again.

Tip: If your horse does not clearly indicate the location of the bag, you can see if this is the case because he does not feel free to take the food or touch the bag. If this is not it and you would prefer that he give a clearer signal to show the bag's location, do not take the bag out of the ground straight away. Instead, wait for him to show the more explicit behaviour you recognize and have been waiting for.

Tip: Be very happy and share this feeling with your horse, but be careful not to express it in a way that can startle him.

Important: When you remove the bag from the ground and your horse really wants it, there could be a bit of a press before you have opened the zipper and he can eat. Do not get mad at your horse for his eagerness. He is still flush with victory and wants the thing for which he has worked so hard. Turning this into a negative moment by controlling or enforcing obedience is not inducive to the tracking.

14.9 HOW YOU CAN TELL THAT THE HORSE HAS LOST THE TRACK

It can be that the horse loses the trail you have walked. He can react to this in a more active and a more passive way. Here are some reactions I see:

BACK TO THE BEGINNING

When he loses the trail, the horse can walk back to the smeller. He could scrape his hoof over the smeller in order to release more scent, and then go back to tracking.

DIFFERENT SEARCH PATTERNS EMERGE

When the horse loses the trail, he can move to a different tracking strategy, one you will recognize if you do treat searches with your horse (Chapter 18). It is a strategy by which he can literally cover a lot of ground, searching it strategically, and it takes less energy than tracking. This search pattern is characterized by the horse moving in large circles and serpentines. The horse does this until he catches the scent he is looking for, which he can then follow through the air. The closer to the source, the stronger the scent (**Fig. 14.44**).

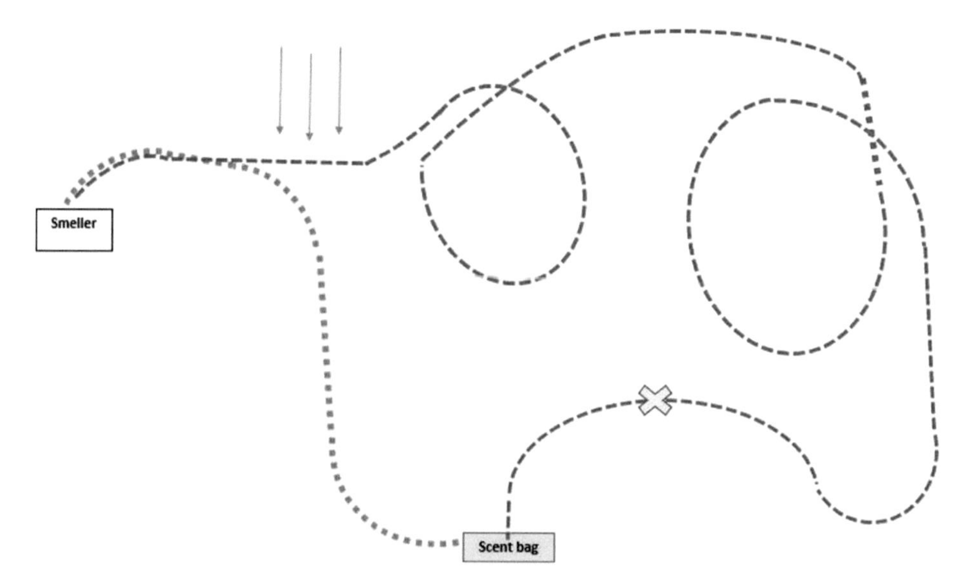

Fig. 14.44 The orange line is the trail walked by the trail layer. The blue line is the route the horse took. The first part of the trail went well, but on the right turn, he went straight ahead instead. He switched to circling right away though, in order to reach the bag. He caught the scent again at the blue cross. The orange arrows indicate the direction of the wind.

CHECKING OLD HIDING PLACES

I cannot comprehend the visual memory of horses; it is that good. Some horses who have lost the trail move to a different tactic. If they are in an arena where they have tracked before, they methodically check the places where they have found scent bags before.

TENSION RISES

When a horse loses the trail for a moment, or when he is generally at a loss for how to find the bag, he can also show calming signals and displacement behaviour. This can include all possible calming signals, but common ones are blinking and immobility. You can also see displacement behaviour, such as when a horse stands still and rubs his head along his foreleg, or bites himself. It could be that your horse heightens his pace, possibly trotting a few steps, or raises his head to the mid-high or high position for short periods, with his eyes opened a little wider. I have not seen real stress signals and body features in response to not finding the bag. You will not see a horse who has to defecate or urinate; who has a significantly contorted nose, mouth, and chin; or who can no longer eat or drink, just to name a few stress-related behaviours and features.

EATING

Eating is a calming signal. It could have been named in the paragraph above, but I am discussing it separately because it is also a specific feature of tracking. A small percentage of horses will start to eat the grass at the edge of the arena, for instance, when they have given up on tracking.

Most horses, however, will eat to give themselves a break in order to calm down. This eating break can take anywhere from 20 seconds to 5 minutes. Afterward, they continue tracking of their own accord, without any pressure or encouragement from the handler. It is as if the eating break was necessary in order to properly continue the trail, perhaps to calm down.

When a horse starts to eat while tracking, a first response from the handler can be to pull the horse's head up and guide him back to the trail because he there here to track not to eat. It feels very counterintuitive when you are so full of anticipation about the tracking to stand there and wait while your horse is eating. Oftentimes the thought is: 'What if my horse likes the grass more than the tracking? Then I am teaching him that the smeller is a precursor to eating grass.' This is a legitimate thought, but it is not what I encounter in the field, but the horses return to tracking.

Can you see the difference between eating as giving up and eating as taking a break?

It is difficult, but here are some indications. The horse who is done tracking is only focused on eating: he will show a 'bite-step' eating pattern, often in a straighter line. The horse who is taking a break can have his hindquarters and flanks turned toward the trail and the smeller. He can sometimes turn an ear or an eye toward the trail (**Figs. 14.45–14.54**).

Figs. 14.45–14.54 Aysa saw the smeller lying in the arena and approached it (Fig. 14.45). When she came close to it, she turned away from it and started to graze (Fig. 14.46). As she grazed, she turned her hindquarters toward the smeller (Fig. 14.47).

Figs. 14.45–14.54 (Continued) She occasionally glanced at the smeller (Fig. 14.48). She also briefly looked at me, then looked away again (Fig. 14.49). She looked left, toward the smeller. Then she turned her whole body to the left so that her flank is turned to the smeller, and every so often, she turned her gaze and nose toward the smeller (Fig. 14.50).

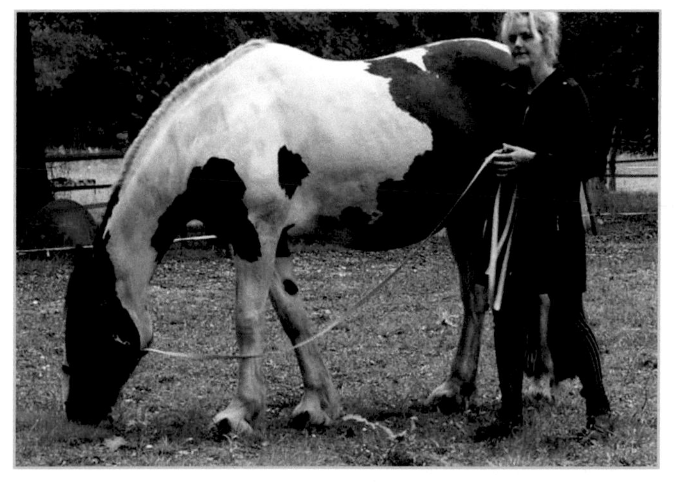

Figs. 14.45–14.54 (Continued) Aysa turned toward the smeller and glanced at it every now and then (Fig. 14.51). She also walked in that direction; however, at the last second, she turned away. She turned right, and from that position, she also bent her head toward the smeller (Fig. 14.52). She continued her turn to the right until she decided to approach the smeller (Fig. 14.53) and to start tracking.

Figs. 14.45–14.54 (Continued) She found the scent bag (Fig. 14.54). The entire sessions lasted 11 minutes.

THE TENSION LOWERS

As indicated above, a horse who has lost the trail can also give up or start doing something else: eating, walking toward the exit, looking at other horses, all without making even one ear or eye movement toward the smeller.

TOUCHING THE HANDLER TO ASK FOR HELP

It is also possible that, if he cannot find the trail anymore, the horse asks for the handler's help. He can touch your hand, bump against you with his nose or head, or stand close to you.

BACKWARDS TRACKING

The horse can also walk back to the trail you laid, smell it, and pick it up again, only now he is following it in the wrong direction, not to the scent bag, but back to the smeller (backwards tracking). So, after he has done the tracking work, he ends up at the smeller, not the bag. Almost all the horses I know turn back around at that point in order to follow the trail in the other direction to the scent bag. As they do this, I no longer see them back tracking; they have gone through an internal learning process that has taught them not to do this again.

What to do if the horse permanently loses the trail

This is the most difficult question to answer because there are a lot of options. It depends on what the horse shows, how he is feeling physically and mentally, what attitude you take during the tracking, what the circumstances of the day are like, what kind of week it has been (for example, does your horse have elevated stress levels from something that happened the day before), how you laid the trail, and so on. In the appendix, you will find a list that might help.

Generally speaking, it is true that you might have to have some patience. Perhaps the horse can still find the bag, and you are being too impatient, expecting him to move at a faster pace. If you really think things are going south, you can lead your horse back to the smeller. Do not do this too often though – a maximum of three times. This is to avoid creating negative emotions in your horse.

If your horse cannot do it, and you think this is because he is not feeling well, or because he is tired, or is too tense, end the tracking for the day and try again another time. The horse can experience walking away from the trail without having found the bag and eating his treats in several ways. If he is not aware of the tracking and the bag full of food that he is missing out on, he will remain neutral, as he does not know what he is missing.

If he is familiar with tracking, and he knows that abandoning the trail means missing out on a fun activity and a delicious food reward, it can be very disappointing and frustrating for him that he has failed to find the bag. If you put the horse back in his stall like this, he can interpret it as a punishment. After all, he is not getting the reward he expected. That is why I look for another bonding activity that the horse and handler can do together so that the joint activity can be ended on a positive note. These activities can include letting the horse graze, giving him hay and standing next to him, or taking a walk together, if your horse likes this. I do not give him the food from the scent bag and no other top five rewards either. That is for when he has done the hard work of tracking. But trying to make a connection and letting him eat something else is a good alternative.

Note: If the horse catches on to how tracking works, there is no way he will find eating grass or hay preferable to tracking and the reward in the bag; he will not choose to do this instead of tracking. This is assuming the horse lives under normal circumstances in which he has enough to eat and drink, other horses around him, etc.

14.10 USING SCENT TRACKING TO ASSESS HOMEOSTASIS

It is common knowledge that horses are very good at masking pain. When moving in a herd, they would be easy prey to a lion looking for a meal if they showed weakness. Masking pain is also something a domesticated horse does, and various studies show that many owners cannot tell when their horse is in pain.[1]

I can imagine that this is less or not true in the case of an acute injury, when the owner sees loss of hair, a wound, a cut, a bruise, a skin abnormality such as lumps or scabs, a thickening, a hardening, or increased heat on a certain area of the horse's body. Things become more difficult, however, when it comes to identifying a toothache, muscle ache, headache, or other aches, such as a recurring light sting when the horse moves his neck to the left; a prickly pain due to arthrosis; cramps caused by colic; or a chronic, nagging stomach ache. And then there are the different stages of fatigue or hormonal imbalance.

You will see that a horse who knows how to track, regardless of the length of the trail, who is no longer learning in any way, loves to do it. He is eager to start tracking and is physically and mentally capable of it. Tracking is more tiring than we often realise. If a horse is not feeling well, for whatever reason, you will often see that he does not start tracking, or if he has started, he is not able to complete a trail he would normally tackle with ease. It could be that the horse makes a passive impression, as if he is too tired. He might smell the smeller, but that is it. You might see calming signals or displacement behaviour.

It can also be the case that the horse is too active to track, too tense to start, and unable to calm himself and concentrate on the job he would normally want to do. You can see calming signals and displacement behaviours, as well as stress signals and features.

Research results that correspond with my practical experiences can be found in Robert Sapolsky's book *Behave: The Biology of Humans at Our Best and Worst*. He explains that depression, pain, anxiety, and stress, among other things, cause the inhibition of dopamine signalling.[2] And we know that the release of dopamine plays an essential role in starting and continuing to track.

If I am in doubt whether a horse I am tracking with is feeling well, then how he tracks is a convenient way for me to determine if he is mentally and physically in balance and in homeostasis.

14.11 HOW OFTEN DO YOU TRAIN?

DURING THE SESSION

When you are tracking and your horse is enjoying it, you can become so enthusiastic that you do too much: laying trail after trail and making them ever longer and more difficult. This can cause your horse to quit and stop tracking during the session. That is a real shame, because you want to end the session on a positive note. That is why it is important to recognize when your horse is getting too physically and mentally tired to achieve a good result in the next session. When I first started tracking, I was mostly focused on studying the way horses approached the smeller. I was hoping to spot enough markers in this to help me predict future success or failure. I was not having a lot of success with it though. The horses almost always approached the smeller with some eagerness. It turned out that the key to recognizing mental and physical fatigue lay in watching how horses walked away *after* tracking. I pay close attention to how a horse holds his head and neck and to his movement pattern as well.

If, after having eaten the contents of the bag, your horse walks away with his head and neck in the mid-high position, and his movements are coordinated and have enough muscle power behind them, it is a signal that you can continue. When a horse becomes tired, you will see that, after he eats the food reward, his head and neck start to droop. You will also see his movements becoming a little more wobbly: the horse loosens a little in his body and movements. He can also walk more slowly. If your horse's head is in the neutral position or below it, it is always an indication to stop (**Figs. 14.55–14.56**, **Figs. 14.57–14.58**). Although, as always, you should look at your individual horse and base your decision on him.

HOW OFTEN TO TRAIN IN A WEEK?

Personally, I have noticed that an average of two to three times a week works well. It allows the lessons learned to take root. For me, an ideal week looks like this: I do an enriched environment or one of the examples discussed in Chapter 1 twice a week. I do scentwork two to three times a week. And I take a walk with my horse twice a week. At the time of writing, Vos is somewhere between 28 and 30 years old. I think these activities keep him physically and mentally fit and bring him joy in life. Regardless of whether you ride dressage, jump, ride Western, take rides in the woods, or if your horse pulls a carriage, it would be ideal if you could track twice a week and do an enriched environment, or a comparable exercise, once a week.

Important: I see scentwork or doing an enriched environment or some other exercise we discussed as a full-fledged main activity for that day. If a horse has done this, I do not schedule a dressage session or something of that nature.

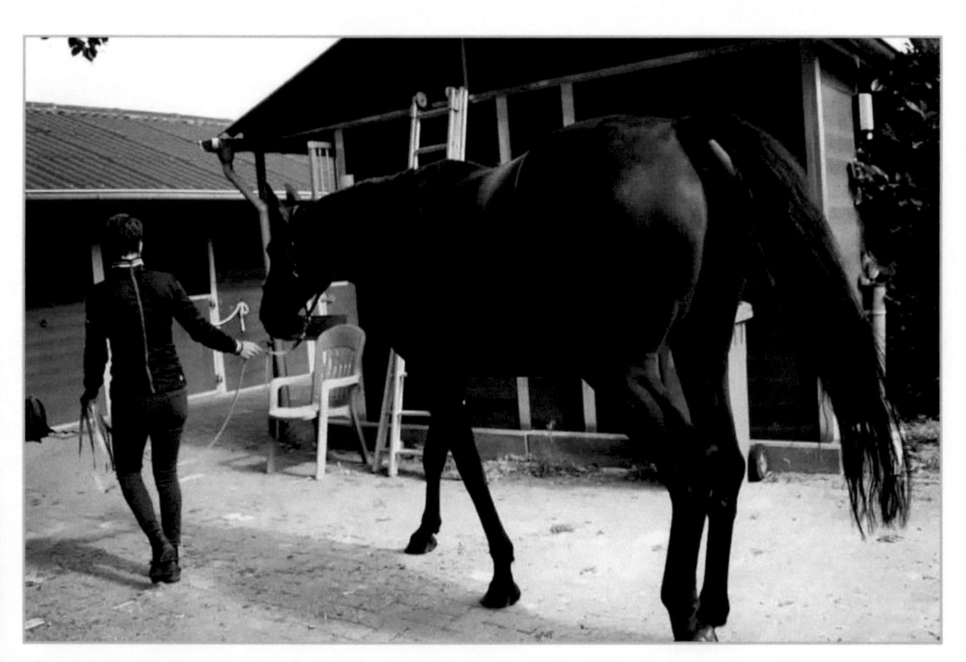

Figs. 14.55–14.56 Picture series of Lieve after her second trail. Her head-neck position was horizontal, and her movements were a little less tight. She was getting tired. We decided to do one more trail, which went well.

Fig. 14.57 Picture of Lieve after her third trail. Her mid-low head-neck position, inward gaze, and her even less tight movements indicated to us that we needed to stop doing tracking exercises for this session.

Fig. 14.58 Picture to compare. Lieve on her way to the arena to track. She had a mid-high to high head-neck position. Ready to go! This does not mean that she always has a head-neck position that is higher than horizontal when she starts. Sometimes she, or any other horse, will already begin tracking outside the arena, on the way to the smeller, and she will have a mid-low to low head-neck position.

REFERENCES

1. Rogers S., et al. (2018) *Equine behaviour in mind-applying behavioural science to the way we keep, work and care for horses*. 5 M Publishing, Sheffield, pp. 160–163.
2. Sapolsky R.M. (2017) *Behave: The biology of humans at our best and worst*. Penguin Random House, pp. 65–74.

15 How to make tracking more difficult

How do you know when tracking is becoming too easy? When your horse completes the trail with such ease, and possibly controlled speed, that he finds the bag in no time (**Fig. 15.1**).

Vos walks into the arena and starts tracking for the bag. He circles left, then right, continuing like this for a long time. I am surprised because I thought the trail was not that hard. Was it too difficult for him after all? I want to look into this. Although Vos is still tracking, I decide, just this once, to lead him out of the arena so that I can lay a new trail. Vos does not let me lead him out though. The moment I make a move to do so, he strides up to the scent bag in a straight line and taps it. How curious. It seems like he was making his tracking last longer before pointing to the bag.

Perhaps I should have payed closer attention in the beginning; maybe Vos had already pointed to the bag with his body posture, perhaps by looking at it or smelling it, and then continued on his way. Because I did not see this, I found myself in the difficult situation of wondering, while he was following the trail, whether it was too difficult for him or too easy.

(I had already excluded the reasons given before in this book for why the tracking might not be succeeding.) And sometimes you cannot avoid having to lay more difficult or easier trails to find out the answer. Laying this trail is a delicate business. When assessing whether your horse needs more of a challenge, you do not want to make it far too easy, but you also do not want to make it so hard that he gives up. Especially since the earlier trail was failure, this would then be the second bad ending in a row. If you do not want to take the risk of your horse failing to complete a trail twice in a row, do not create a more difficult trail, but an easier one instead, and see if your horse is able to finish it.

Fig. 15.1 Vos trotted from the smeller to the bag, finding it without difficulty. It was time to make things a bit more difficult.

15.1 IDEAS TO MAKE THE TRACKING MORE CHALLENGING

If your horse is ready for a challenge, you can expand the tracking in the following ways:

- You still wear the same 'trail shoes', but you create shallower footsteps in the ground when you are laying the trail.
- You can take some sharper turns. If you later start tracking in an area that has a lot of trees, you can also take sharper turns there.
- You can make an old trail and a new trail, creating a scent distraction for your horse. The horse finds the bag at the end of the new trail.
- You can have other people cross through your trail, causing a scent distraction for your horse.
- You can place objects inside the arena. Your trail can go over them if you put down a large cloth, towels, or a few poles. Or your horse can walk around objects when you place a larger obstacle in the arena, such as a large, sealed bale of hay.
- You can increase the time between laying the trail and letting your horse track. Do this by intervals of only a few minutes, and keep the weather and any other circumstances in mind so that the horse still has a fair chance of following the trail.
- You can expand the location in which you lay the trail. For instance, you can place the smeller outside the arena instead of inside it and then make the trail lead into the arena. You can also begin your trails in different locations.
- You can lay your trails in different locations, on different terrains, and on different surfaces. A stone surface is the most difficult

Some examples (**Figs. 15.2–15.11**):
Different location (**Figs. 15.2–15.4**):

Fig. 15.2

Fig. 15.3

Fig. 15.4

Tracking across poles
Fig. 15.5

Fig. 15.5

A track across towels:
Figs. 15.6–15.11

Figs. 15.6–15.11

Figs. 15.6–15.11 (Continued)

A track across towels.

Figs. 15.6–15.11 (Continued) A track across towels.

DIFFERENT TRAIL LAYERS

You can also see if your horse will follow a trail laid by a different person than the one he is used to. This change can be more difficult for some horses than others. If your horse does not follow the scent trail of the new trail layer, begin the teaching process again from the start: bury the bag very close to the smeller, have your new scent layer step up to the bag, and expand the distance when your horse has good runs. Because your horse knows how to follow footstep trails already, you will see that, after you repeat the first steps, you will quickly be able to make large strides in distance and time.

If you want to look for missing persons with your horse, this is a good preparation for it. If, at a later point, you are planning to take the role of missing person in which your horse tracks you, it is also useful to accustom your horse to a different handler. It could be that this is no problem, if, for example, your horse is accustomed to being cared for by different people. If it is a problem, you can try the activities described in section 14.1, which improve the relationship between handler and horse.

SPLIT TRAIL

Laying a split trail is another way to add a challenge for your horse. For a split trail, you walk arm in arm with another person for a time, and then you split apart. It is important to decide beforehand which trail layer will lead to the buried treasure, you or the other person, because your horse will be curious about this new scent. The one whose trail leads to the scent bag leaves his scent print behind on the smeller. After all, the scent on the smeller is the one the horse is supposed to follow. The following example is set out in the arena. You can, however, do this in various locations (**Figs. 15.12–15.19**).

Fig. 15.12 Two trail layers. The first one (Marian) laid the trail Vos is supposed to follow. She was the one to leave her scent on the smeller.

Fig. 15.13 The two trail layers set out together, walking arm in arm or just very close together.

Figs. 15.14–15.15 The two trail layers split up and continue walking.

Figs. 15.14–15.15 (Continued) The two trail layers split up and continue walking.

Fig. 15.16 The trail layers continue their routes.

Figs. 15.17–15.18 The trail layer who stepped on the smeller buries the bag. The mesh is uncovered and open to the air.

Fig. 15.19 Both trail layers leave the arena. Time to collect the horse and see what will happen.

STAY LOGICAL AND UNDERSTANDABLE

For all expansions, the goal should not be to stump your horse, causing him to fail, but to balance the difficulty in such a way that it is harder but still doable. That is why you should try to keep your trails logical and understandable. For example, if your horse is about to follow someone else's trail for the first time, maintain the same rules as before. Have the new trail layer put on old shoes with deep grooves.

If you are leaving time between laying the trail and the tracking, adapt this to the circumstances of the day. If there is a strong wind, it might be good to leave less time in between than if there is no wind. If you are going to place objects in the arena to lay a trail across them, it is only fair for your horse to have seen these objects before in an enriched environment. And if your horse does not cross over the object, but walks around it, picking the trail back up on the other side, let him. He is still the one calling the shots

It is true for all expansions that your horse might have to get used to it. What seems logical to us, might not seem logical to him. It can also be the case that he is not feeling well that day, is too tired or tense, or is in pain.

If the tracking challenge you have devised is not a success, go back a few steps (figuratively) to try to find out where the problem is and work on that. One of the most difficult parts of tracking is that we, as handlers, are challenged to really look at our horses, think about what we are asking them to do, see if they can do it, and change things if necessary. As a behaviour consultant, going through this process is always a learning experience for me. And it triggers our own 'seeking mechanism' too, constantly making us search for new challenges and solutions.

Note: It is nice for a horse if, once in a while, he has a lucky break, and he is not always pushed to the limits of his ability. If you keep making every exercise more difficult, the horse may give up at some point. So play with this a little; make an exercise more difficult every once in a while, but not always.

16 *Scent tracking while riding*

The horses with whom we track while riding, according to my methodology, all started tracking with the handler walking beside them. They are horses who, besides tracking, were also being ridden, be it hunt seat, dressage, recreationally, or Western style.

When these horses are easily able to follow a trail with their handlers walking beside them, are physically capable of doing so, and able to concentrate on it for longer periods (when distracted, they might raise their heads briefly before continuing with the trail), then you can choose to have the horse track while being ridden (**Figs. 16.1, 16.2**).

Fig. 16.1

This has advantages, especially in areas that are suitable to the laying of long trails. That way, the horse can cover greater distances than the handler would be able to on foot, and the horse can also carry a pack (perhaps containing items needed by the person he is looking for).

For a horse who is already a good tracker when his handler walks beside him and who has already been ridden, there are no special transition requirements when he starts tracking while being ridden. In terms of tracking logic, there is no real difference. What you will see, however, are associations the horse and rider have formed while riding. For instance, a horse who is ridden with a bit in his mouth might be afraid to stretch his neck down too far, because he expects pain in his mouth because for tracking, he has to bend down further than the 'normal' neck extension while riding. It can also happen that the horse is used to being strictly controlled while ridden. The freedom he is allowed in tracking can cause extra excitement. Or, contrarily, the horse might have to overcome a tiny barrier to take

Fig. 16.2

the same 'tracking freedom' while being ridden.

When the horse is used to being ridden with high intensity and tension, it can take longer before he is able to calm himself enough to track while being ridden, a problem he had to a far lesser extent while his handler walked beside him during tracking. For the rider, it can be unusual to have to give the horse that much leeway in the front. In order to track, the horse sometimes has a deep posture, taking up all of the reigns. In that case, the rider is literally left empty-handed sometimes and that can be a little scary, also because the deep head-neck position of your horse gives you a different view of his head and neck – more than ever, you are looking at nothing. Add to this the fact that the horse uses his body fully while walking (see the chapter on biomechanics), causing a stronger swinging motion in his back.

Important: When you start having your horse track while riding, you need to lengthen your reigns. I have noticed that the standard dressage and jumping reigns are too short. You can order extra-long reigns, or you can add a section to the existing reigns. Especially when you are riding, you want to maintain a good balance and not cause additional physical strain for your horse by having to lean forward because your reigns are too short.

17 *From finding a scent bag to finding missing persons*

As DISCUSSED in Chapter 11, searching for missing persons is the fourth step in my methodology. I do not progress to this step until the horse is able to track scent bags (while his handler rides him or walks alongside him), can complete a trail that is between 10 and 30 minutes old, and can concentrate for at least somewhere in the neighbourhood of 8 minutes. The goal of tracking for missing persons is finding someone on the basis of that person's scent, which he has smelled on the smeller.

For this exercise, the chosen missing person carries the smeller and scent bag. She lays down the smeller in a previously agreed upon location and stomps and wipes her feet on the smeller a few times. That way, the smeller carries the scent of her shoes and possibly her body odour, if she carried the cloth with her. Then the missing person lays a trail she knows the horse can handle and hides at the end of the trail. At an agreed upon time, or after a phone call from the missing person, horse and handler set out. They walk to the smeller. The horse follows the trail and finds the missing person and the amply filled scent bag and is allowed to eat its contents. If your horse does not find the missing person and the bag, all the challenges and tips that have already been discussed can be relevant. Brainstorm what might have caused or contributed to your horse's failure to find the person, and then try again on the same or a different day with adjustments.

You can imagine that when you start tracking for missing persons, there are a number of differences with tracking for a scent bag in a small location. There, if you just hide the scent bag, adding nothing else, the environment can be pretty 'bare'. If a person is hiding, by definition, there have to be plenty of obstacles, bushes, and trees behind which she can do so.

It might also take a bit more preparation and organisational talent. If you take the trailer to the location, where do you park and lay down the smeller? How long does it take for the missing person to lay a good trail and you can start tracking? Does the time between the laying of the track and the actual tracking, match the time and skills the horse can handle? When you are walking or riding your horse to the smeller, you have to be aware of whether the horse can see (or hear) the missing person hiding herself, so that he already knows where to find her before the tracking even begins. This means you might have to bring an additional handler to guard the smeller while the missing person lays the trail and the horse and handler have not yet arrived. That way, you prevent things like a passing dog snatching up the smeller.

Also check beforehand to see when there is more distraction, in the form of dogs, hikers, cyclists, game, or other stimuli at the location (or maybe you want to purposefully *not* do this!). Use this knowledge to lay a customised trail, and for socialisation. If your horse has more trouble adjusting to the location than you had thought he would, make the trail a little easier, and do not start at the level of difficulty he was able to handle in the old location. Build this back up slowly. The more you align the tracking of missing persons with the tracking the horse has done in the past, the easier the transition will be.

Important: In most places I visit where horses are stabled, forests or other natural areas are not easily accessible. If you live in a place where you do have easy access to these places, you can also do the first three steps of my methodology in the forest or natural area right away, gradually building up all the skills a horse needs to find missing persons in that location.

17.1 ADDITIONAL SOCIALISATION TO 'MISSING' PERSONS

As noted in Chapter 8, horses are quicker to show a stop and flight response than dogs when they see, hear, and smell people who are hiding. It is amazing how powerful this evolutionary mechanism is, how it allowed the horse to survive during his entire long history. Even if your horse knows someone well, if that person suddenly hides, does not make a sound, and stands or squats behind a tree or bush, the horse can respond differently than you would expect (**Figs. 17.1–17.8**).

Figs. 17.1–17.8 Indy started tracking. I was sitting behind the bush in the foreground of the picture. Indy tracked, but did spot quickly enough that there was something different about this bush. This makes sense, I think because, normally when I am with her and her owner, Ristin, I am not squatting behind a bush.

Figs. 17.1–17.8 (Continued) She alternated between tracking and looking at me.

Figs. 17.1–17.8 (Continued) Then found the scent bag in front of me.

If your horse is not yet fully socialised to possible 'missing and hidden' people, it would be good to do some preparatory work. After all, after your horse completes a difficult tracking run and manages to find the bag with the missing person, you do not want him to have such a negative experience finding the person that it negatively affects the tracking exercise (both retroactively and for the future). You can have your horse get used to hidden missing persons by doing the following things yourself first.

OPTION 1

Stand, squat, or sit in different places near your horse. You can sit on a chair or small stool, and later on the ground. If your horse is at pasture, you can do this from the other side of the fence. When you are taking your horse for a walk and there is a good place to sit (e.g., a log or a bench) you can sit down while your horse is grazing or is enjoying the view. When your horse is eating hay, you can also stand, squat, or sit beside him.

Important: Safety first. Make sure you maintain enough distance from your horse to keep him in his comfort zone – the horse always has to be able to walk away from you, choosing the distance that he finds comfortable. You do not want him to startle and thereby cause an unsafe situation for you. You can gradually stand, squat, or sit closer and closer to your horse. Once your horse is fine with you standing, squatting, or sitting close to him, whether on a stool or not, you can also gradually start to introduce other people. That way, he gets used to people with different looks, clothes, hats, movement patterns, voices, and scents. But remember, it is better to be safe than sorry. It is possible to let your horse get used to other people when they passively stand, sit, and squat on the other side of the fence. This works well together with the second, following option

OPTION 2

You let the horse find the scent bag, as you normally would; however, behind the scent bag, on the line along which the trail would have continued, you have someone stand, squat, or sit. The distance this person has to the scent bag depends on how apprehensive you expect your horse to be about his presence. And again, it is better to be safe than sorry. If you are not sure how your horse will react to this person, practice the previous step a little more, or increase the distance between the scent bag and the person. The missing person, who is present at a certain distance, is calm and does not move. She does not suddenly stand up when the horse finds the bag. She is just present; that is all. The only movements she makes are calming signals, like turning away the head or shoulders. The moment the horse becomes more comfortable with the missing person, the person can come a little closer to the scent bag, and she can hide a bit more later as well.

Note: During my tracking lessons, I am the one who lays the trails, while both horse and owner do not look. I also play the missing person in the forest, so the horse follows a scent trail with which he is already familiar (**Figs. 17.9–17.13**).

Figs. 17.9–17.13 I laid down the smeller in the woods. Indy set out to find me. She did this with great care, continuously checking to the left and the right of the trail to make sure she was still following it. The trail went through the forest, up a hill, and across a fallen log. It took Indy about 40 minutes to find the scent bag and me. Ristin, the handler, detached Indy's rope in a denser thicket so the rope would not impede Indy's progress.

Figs. 17.9–17.13 (Continued) I buried the scent bag 1.5 metres in front of where I was sitting.

Fig. 17.13 The view from my perspective as Indy and Ristin were approaching me.

OPTION 3

You can have your horse approach certain people without the association of the smeller. Do or do not use food for this, depending on your own insight.

- Step 1: A missing person shows the horse an apple and lets him smell it, and then she walks backward.
- Step 2: The horse is held back by the handler, and then he is allowed to follow after the missing person.

  Important: There is a delicate balance to find: for a second, you are holding your horse back, and then he is allowed to go. Ideally, you hold your horse back with no or just a little bit of tension on the rope, while the horse is still occupied with the person holding the food who is walking away. Do not restrain your horse forcefully. If your horse is eager to follow, let him. Better too soon than too late, even if the person holding the food is not yet in position. Otherwise the horse can begin to interpret being held back as a command not to follow the person.
- Step 3: The horse follows the person holding the apple and gets the reward when he reaches her.
- Step 4: The person holding the apple can sit or kneel and change positions.

17.2 HAVE MISSING PERSONS GIVE CALMING SIGNALS

If necessary, the missing person can give calming signals to the horse at the moment the horse sees her and approaches her. This can be one, two, or three signals, with a little time in between. This gives the horse that little bit of reassurance as he approaches the person. By giving the calming signal, the person is telling the horse that she does not seek conflict and that she wants to start or maintain a positive social relationship.

For instance, missing persons can look away, making a head turn away from the horse. They can also look just to the side of the horse instead of straight at him. They can turn their shoulders away diagonally, stand or sit with their sides facing the horse, or lower their chins to their chests a little. If the horse is comfortable with the missing person, one calming signal, or none at all, will be fine. Perhaps I do not need to say this, but it is fine for the missing person to talk to the horse when the horse approaches. That way, the horse can recognize the person by the sound of their voice, but the missing person should not make sudden movements, as this can, of course, startle the horse (**Figs. 17.14–17.18**).

Figs. 17.14–17.18 Sybrand showed Vosje the treats he had for him (Fig. 17.14). Vosje was definitely interested.

Figs. 17.14–17.18 (Continued) Sybrand walked away and I held back Vosje, who wanted to follow (Fig. 17.15). Vosje walked toward Sybrand, but in a big arc (a calming signal) and then approached Sybrand. Sybrand made a head turn as a calming signal (Fig. 17.16).

Figs. 17.14–17.18 (Continued) Vosje ate the treats (Fig. 17.17). Sybrand stood up and gave Vosje some additional treats. Vosje checked to see if the container was really empty (Fig. 17.18). (Note: Prior to this session, we had a similar session with Vosje and Sybrand. Then, Sybrand was standing while Vosje approached him, and Vosje moved toward Sybrand in a straight line).

17.3 WHO OPENS THE BAG?

The choice of who opens the bag once the horse has found the missing person, the owner or the missing person, depends on the horse's comfort level around the missing person. If the horse is entirely comfortable, either person can do it. If the horse still has to get used to the missing person, even just a little, I have the owner open and hold the bag while the horse eats. That way, I make sure the horse can eat the reward without worry, while the missing person calmly stands close to the horse, so he can get used to her while experiencing a positive emotion.

17.4 FURTHER DEVELOPMENT

On location in the natural area, you could add more distance and make your trails for missing persons longer. You can also use the ideas I shared in section 15.1.

Additionally, my methodology is predicated on having the horse use his nose in an effective and functional way, but you will see that, once a missing person comes into play, the horse also very clearly uses his sight and hearing. At this point in the learning process, that is no problem at all, because he has already mastered using his nose to track. This allows him to use all of his senses optimally, between which he will alternate or which he will use all at once. For long trails, he might start off using his nose, to which he will later add sight. This is no problem. The goal is for the horse to complete the task using the senses he chooses. Every missing person found is cause for celebration.

Photo series: Vosje and Imke (**Figs. 17.19–17.21, Figs. 17.22–17.29**).

Figs. 17.19–17.21 Imke sat on the wooden stool to have Vosje get used to her. Vosje came over and Imke started feeding Vosje some handfuls of grass (Note: the electic fence is off).

Figs. 17.19–17.21 (Continued) Imke sat on the wooden stool to have Vosje get used to her. Vosje came over and Imke started feeding Vosje some handfuls of grass (Note: the electic fence is off). We proceeded. Imke made a small scent track and sat behind a tree, out of sight of Vosje. I collected Vosje.

Figs. 17.22–17.29
Imke laid the track and was sitting behind the tree waiting for me to collect Vosje (Fig. 17.22). In Fig. 17.23, I approached the smeller. Vosje took a sniff, and bent to the right to start eating grass (Fig. 17.24).

Figs. 17.22–17.29 (Continued) After a few seconds, Vosje proceeded to walk toward Imke (Figs. 17.25–17.26). When closing in on her, he dropped his head to take another bite of grass (Fig. 17.27).

Figs. 17.22–17.29 (Continued) Imke opened the bag though, and Vosje proceeded to the bag (Fig. 17.28). Imke slowly got up and stood up straight while holding the scent bag for Vosje to eat from (Fig. 17.29).

18 *Treat search*

TREAT SEARCH: *The* instrument that brings optimal value, even though it takes the handler very little time and effort. I learned this activity and the many benefits it brings in Turid Rugaas's training course. It works just as well for horses as it does for dogs, and almost all the benefits of scent tracking can also be had through a treat search. A treat search, as the name says, is about your horse searching for all sorts of treats that have been scattered around. These can be small piles of hay, feed pellets, or fruits or vegetables that you have cut into small enough pieces beforehand. You scatter the treats in such a way that the horse has to put in a little effort to find them but that he is capable of finding them, which makes him motivated to continue searching and, in the end, gather a nice meal for himself.

The treats can be scattered in a space that is familiar to the horse, such as the pasture, paddock, or stable aisles. You can also bring the treats on a trip or on a walk. You can scatter them beforehand and then bring your horse so he can find them or scatter them as you are walking together.

With most horses, you will not meet any barriers when you do a treat search, and you can follow these steps.

- Step 1: You stand by your horse, let him smell the pieces of food, and throw them onto the ground a metre in front of your horse, in an area of about a square metre. Then you stand very still. Perhaps you look at the bits of food, but you do not push the horse, point, or anything like that. Normally, the horse will start to collect the pieces of food, which are still relatively close together.
- Step 2: Do the same thing as in Step 1, except this time, hold back a few pieces of food and do not scatter them inside the square metre but just outside it. That way, you increase the area in which the horse needs to look.
- Step 3: Gradually enlarge the space in which you scatter the bits of food and the distance between the pieces, but make sure it is still doable and the horse has a quick rhythm in searching for and finding the pieces.
- Step 4: If your horse knows exactly what he has to do and starts off enthusiastically, you can begin marking the start of the search with an arm gesture or a spoken prompt, such as, 'go have a look,' or 'where is the food?' Personally, I do not specifically use the word 'search' because the horse shows a very different walking pattern during a treat search than when he is scent tracking (**Fig. 18.1**).

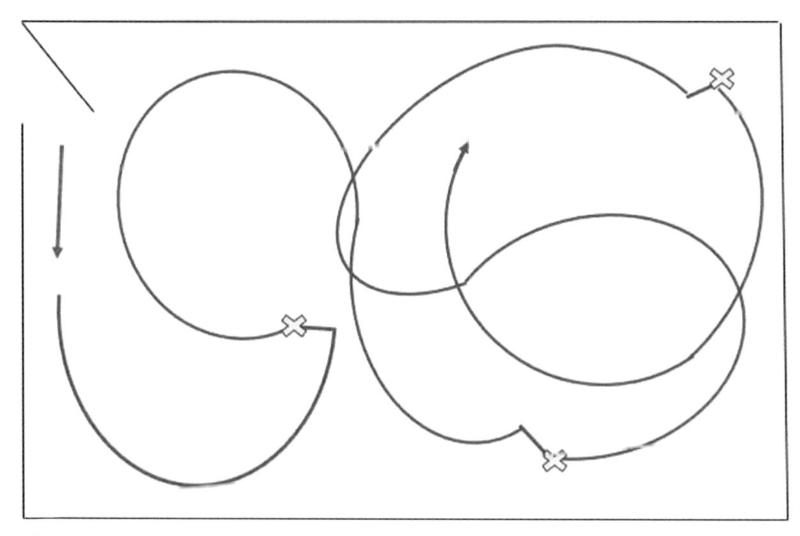

Fig. 18.1 The walking pattern during a treat search.

As you can see, the horse walks a lot of circles and serpentines. I think this is the natural movement pattern of a horse who is searching and has no scent clues to go on. This way, the horse covers a lot of square metres in an efficient way. When he gets close to a piece of food, he catches its scent and is able to find it. That is why, as he steadily walks his serpentines, he can stop suddenly every once in a while when he smells the hidden food or something else (at the crosses).

DO NOT USE THE 'SEARCH' COMMAND FOR BOTH DISCIPLINES

Back to the search prompt. The word 'search' is on the tip of many people's tongues when they want to stimulate their horses to start searching. If you are not doing scent tracking with your horse, this is no problem. However, if your horse knows the word 'search' from doing treat searches, and you then use it for tracking, you will see that, in the first few tracking sessions, when you get to the smeller and say 'search', your horse will immediately move away from the smeller and start to walk in circular patterns. During the later sessions, when your horse has a handle on tracking, the chance of this is much smaller. To avoid confusion, I advise you to use a different term for each discipline if you do both, or just use a specific term for treat search and none at all for scent tracking. After all, seeing the smeller is already a sign to the horse that the tracking has begun.

Actually, why do I use a search prompt or a specific gesture for treat search at all? It is possible that I have scattered food bits in a pasture for one or more horses and they do not see it. They could just be coming from their stalls or changing pastures. Because they did not see me scatter the pieces of food, I show them my arm gesture and say, 'go have a look'. Then they will know there are things to be found, and they will start to search. The search prompt is not necessary because the horses often find the food bits anyway, as once they have come across one, they usually start searching for more. Because I often like to watch their search, I give them a prompt, so I do not have to wait for them to find a piece of food accidentally, which can take some time.

You can expand the size of the treat search, and you will see your horse improving when you do this, just like with tracking. After some time, it is quite normal for your horse to be able to search for 20–40 minutes. However, this is only the case if he is looking for food bits he really likes and badly wants to find.

A SANDY SURFACE

When you scatter your treats on sandy soil, there is a good chance that the sand will stick to any pieces of fruit you have scattered, or that he will ingest sand when he eats the treats. This is unhealthy for the horse, especially if you repeat this activity often. In order to prevent this from happening, you could look for places without sand. If these are not available, you can also work with multiple scent bags or containers with holes in them. I buy these at fish and tackle shops. They are containers meant for holding live bait. Of course, you can also make your own containers, using paper boxes or bags. It is easy to poke holes in the tops of these; however, when the ground is wet or moist, these boxes quickly disintegrate. You can also use a small cloth bag and poke holes in that (any small bag will do, as long as sand cannot get into it). The bag or container does not need to be as large as the scent bag. The horse will search for multiple rewards, and once he finds them, you take out the food, and then he will continue looking for the next small bag or container. Together, these small rewards form the whole meal. Here again it is important to use treats your horse really likes. And use different ones; that way, the search stays exciting for your horse.

Tip: If you intend to do treat search using containers or bags, it is best to teach your horse how to do it in a pasture, a grooming area, or another place where there is not a lot of loose sand. Once your horse understands what he has to do, and he knows the arm gesture and verbal prompt, you move the search to the sandy area and start using containers or bags. You fill these containers with food and then you bury them. You do this in such a way that the part with the holes in it is above the surface. That way, the horse does not see the container, but he does smell it. Then, when he starts searching on the sandy soil and smells the food, he will touch the container, pick it up, or flip it out of the ground. That is when you take over and open the container for him.

Important: If you have a horse who circles back and forth a lot, or a horse who will bite the container in order to get the food out, do not use hard plastic containers. There is a good chance that they will crack into sharp plastic shards. Which you want to avoid of course.

Some examples:

(**Fig. 18.2, Fig. 18.3, Fig. 18.4, Fig. 18.5, Fig. 18.6, Fig. 18.7, Fig. 18.8, Fig. 18.9, Fig. 18.10, Fig. 18.11, Figs. 18.12–Fig. 18.17, Fig. 18.18**)

Fig. 18.2

Fig. 18.3

Fig. 18.4

Fig. 18.5

Fig. 18.6

Fig. 18.7

Fig. 18.8

Fig. 18.9

Fig. 18.10

Fig. 18.11

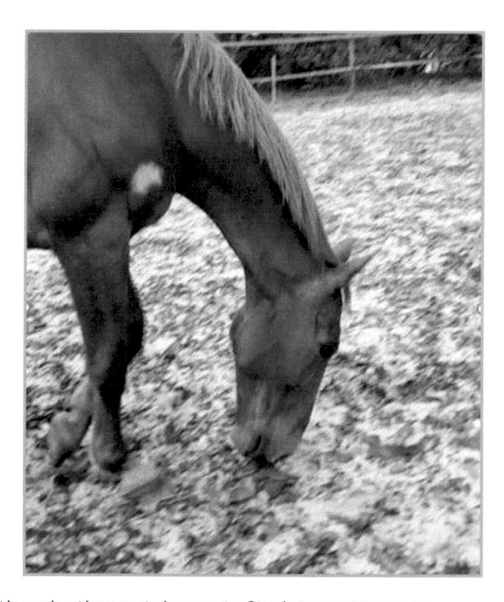

Figs. 18.12–18.17 An example of a treat search using the plastic containers. Left picture: Vos was looking for food containers. He was close to one of them. Right picture: There it is! His nose went far down to the ground for a second. I could not see the container.

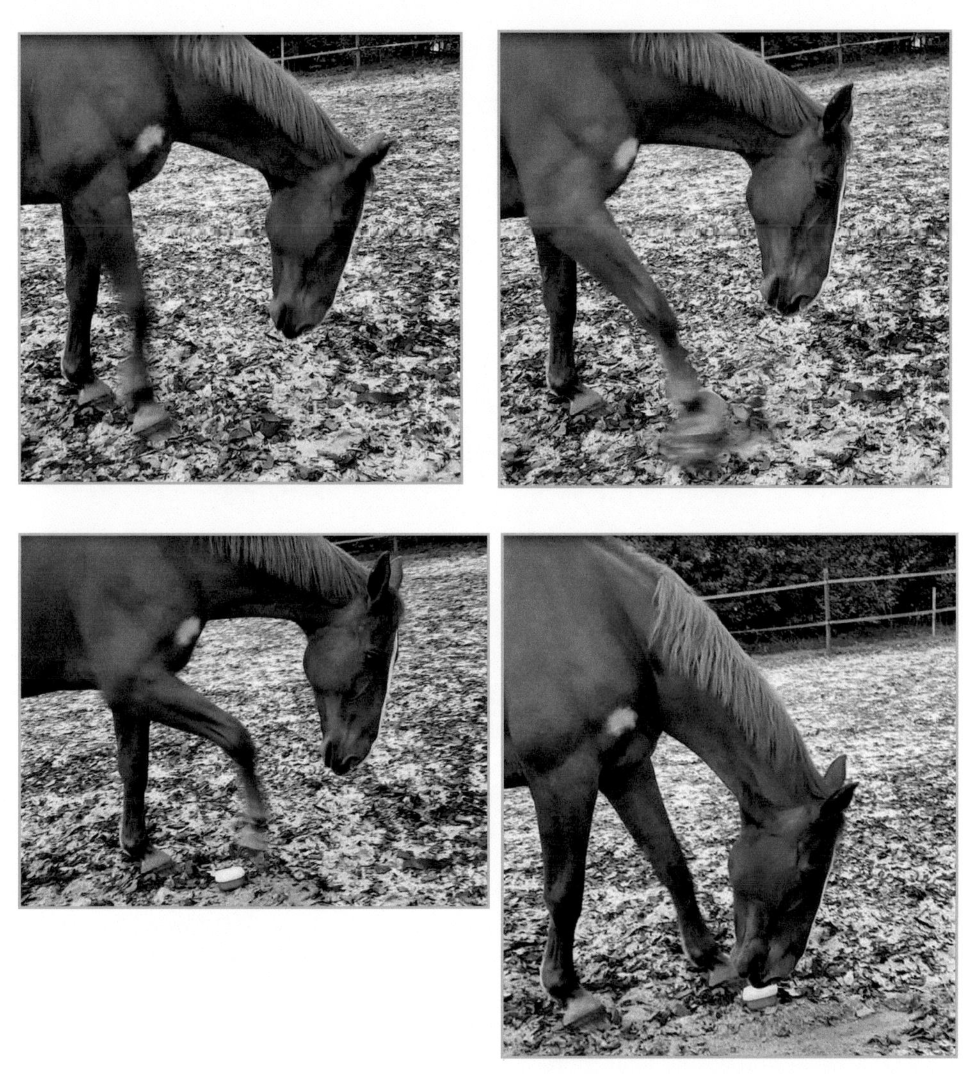

Figs. 18.12–18.17 (Continued) Upper left picture: Vos leans forward a little, raising his head. Upper right picture and bottom pictures: He scoops the container out of the soil and taps it with his mouth.

Fig. 18.18 Use a system to remember where you hid the containers. If your horse is unable to find them, you will have to retrieve them yourself. This is a good way to learn how frustrating and annoying it is when you cannot find an object. In inset photo, Vos and I were searching together. In the larger picture, he had already given up.

ABOUT COLOUR

A horse sees colours differently than we do. When you see comparison images of how horses and humans see the world, in the eyes of the horse, everything seems to consist of a lot of shades of grey and green. The colours in the orange and red range seem to transform into greens, and colours that contrast with that are blues and yellows.[1] This means a piece of orange carrot or apricot is hard to see in the grass, but a yellow banana with peel is more visible.

ALONE OR TOGETHER

You can have your horse do the treat search on his own, but it is also possible to have a group of horses do it together. After all, searching for and eating food together is a social activity. You have to be able to do it safely though. The horses I did this with were suited to it. They were in groups that had lived together for a longer time and none of them have shown food aggression toward one another. They also did not show inappropriate behaviour toward me by all rushing at the bucket in which I carried the food bits. If you are not sure about the horses' behaviour, do not take any risks, and slowly build toward this.

You can achieve this by doing individual treat searches with each of the horses first and then doing the group treat search so that each individual horse knows what the activity is about. That way, they will already have experienced the benefits if searching: the fun of exploration, the dopamine rush, the calming effect. By that point, they will already be able to concentrate for longer periods or maintain the required body posture for longer. That makes the start with the group easier later on.

If I scatter food bits among a group of horses, I make sure that I have enough treats with me from the start, and I scatter a good handful for each horse. That way, they all have enough to eat right away. In addition, I scatter many more treats on the ground in between the horses. I do this to give them the impression right away that there is enough for everybody, and they will already have eaten a portion before they continue to search. Here, it is also the case that you can increase the distance between the treats as time goes on. If you are walking through the pasture with a bucket of treats, it could be that a horse prefers to approach the bucket to eat rather than search. In that case, I do not tussle with him to keep him away from the bucket. I give him some treats, and I also scatter some. That way, I keep the horse in a relaxed state, and he will still start searching after all. After six or eight times, you will see that these horses no longer approach the bucket but start searching right away and keep doing so. The fun of the search takes over at that point. Of course, if the circumstances allow for it, you can also toss treats into the pasture from the other side of the fence, or scatter treats when the horses are not at pasture yet.

ON AND OFF THE ROPE

During a treat search in which you do not use containers but scatter food bits on the ground directly, you can release your horse into a safe, enclosed area, thereby giving him a nice mental and physical exercise while you do something else. You can also see it as an activity you do together, in which you walk alongside your horse, or even in which you both search, and you call him over when you have found something if he can no longer find anything himself. You can also do a treat search with your horse outside the enclosed areas, for instance by hiding treats in different places, such as in the parking lot, in the woods, on the path, etc. It can be any place you think is suitable. Perhaps, because of sloping surfaces in different places, this is an additional physical exercise for your horse. Or perhaps you want to practice a bit closer to certain stimuli. It goes without saying that, in these cases, you hold your horse on a long rope and follow alongside him just as you would when you are tracking. It is also a nice activity to strengthen the bond between you and your horse. Here, too, there can be reciprocity if you find a carrot (possibly calling your horse over) and give it to him. An additional advantage is that, by searching together, you can show your horse that he can take initiative and walk away while you are close by.

START-UP PROBLEMS

A small percentage of horses have trouble with the treat search at first. The moment you scatter bits of food in front of them, they continue to sniff your hand or stand next to you – they do not move down to the ground to look for treats. It is as if you have a pulled a disappearing act, and the horse does not know where the apple pieces went. The horse seems to have made a strong link between the person and the food. If this person is here, the food must be coming from the person: there is no other way! It can also be the case that the horse has learned that he cannot take initiative beyond what he has been taught to do when there is a person present.

If you encounter this challenge, always find out first if the horse will start to search for the food if the owner is not standing close by. You do this to see if he uses his nose, or possibly if there is a physical problem. I do this by scattering various treats within a space of a square metre, at a distance of a metre and a half in front of the horse, and then walking away and looking from a distance to see what the horse does and whether he finds the bits of food. With the domesticated horses I work with, I have never seen a horse not find and eat the pieces of food. If your horse does not search for food, there are several ways in which you can solve this. I will name a few. The first one on this list is the most natural to me and is my preference.

SHARING FEEDING TIME

You stand next to you horse more often when he is grazing at pasture or eating in his stall. That way, he can get used to you being there while he eats. You can take him for walks a bit more often, allowing him to graze in places where it is safe to do this. Once your horse has gotten used to you being there during these eating sessions, you can put some treats in with his food while you stay there with him. You can also do the haystack exercise.

HAVING SOMEONE ELSE SCATTER THE FOOD

You have your horse on a long rope or in an enclosed space. You do not make contact with your horse and stand with your flank turned to him. Have someone else scatter the food right in front of the horse. You do not look at the scattered food, and you try to maintain your invisible pose as the horse starts to search and eat. As time goes on, you turn a bit more toward your horse or move a little closer to him every session. You do this until he is fine with it, knowing he can search for bits of food with you there, and you can scatter them yourself.

LOWER YOUR HAND TO THE GROUND

You go through several stages in which you teach your horse he can eat food bits off the ground. You do this by incrementally, with every reward you give, lowering your hand closer to the ground. At a certain point, you will be squatting and your hand will be on the ground, with your horse eating the apple (or other reward) out of it. If this works, you take multiple apple pieces in your hand, have the horse eat them off it while your hand is on the ground again, and let some pieces fall off it, onto the floor. As your horse eats them, you stay squatting next to him. Step by step, you give your horse more and more food rewards on the ground next to your hand, until you are no longer using your hand but are still squatting. If this goes well, you can gradually leave your squatting position, so that after a few sessions you no longer have to squat at all when you scatter the bits of food (if you get up too quickly in the beginning, your horse will bring up his head and neck along with you).

18.1 WHAT TO DO FIRST: TRACKING OR TREAT SEARCH?

As you have seen, tracking and treat search have a lot of things in common, but there are also differences. If you have done neither with your horse yet, I would not start doing both at the same time, as this may confuse your horse. If you do intend to do both in the end, I would start with one. Once your horse has made a good leap forward in that discipline, attainting basic proficiency, such as being able to follow a 6-metre scent trail eight out of ten times or being able to look for bits of food for 5 minutes in a treat search, then you can start doing the other discipline as well, at which point you can do both at the same time.

I do not have a preference for which discipline you should do first. It depends on what you think would best suit you and your horse. It also depends on the things you have already practiced with your horse. I am assuming that, in preparation, you regularly take your horse for walks and that you can take various positions around him when you do so. I am also assuming that your horse has no trouble walking away from you and taking initiative when you are there.

If you start doing treat search first, it has the advantages that your horse already practices using his nose a little, holding the tracking posture, and concentrating. This can make the start of the scent trail easier. The disadvantage of beginning with treat search can be that when the horse uses his nose, he automatically goes into a circling pattern, because this is a little less intensive than following the scent on the trail. In order to prevent circling, you will have to pay a little more attention to the start of scent tracking and lengthen the trail in very small increments to prevent your horse from being tempted to start circling right away.

If you start with tracking, and your horse has the basic proficiency that he can nearly always complete a trail of 6 to 8 metres, it can be confusing for him when you start doing treat search. There is no white cloth and no trail for him to follow. The horse will move to circling and air scenting fairly quickly, however. Then when you start tracking again, you will see that he tends to begin circling and air scenting less quickly, and that he picks up the scent of the smeller easily.

Important: In treat search, you want your horse to be able to find a large quantity of food in a relatively short time; however, it should not be the case that he does harder work than he would on a scent trail and gets a smaller reward for it. This is demoralising in both disciplines.

NOTE

1. University of Exeter. (2018) *Research into equine vision leads to trial of new fence and hurdle design to further improve safety in jump racing.* https://www.exeter.ac.uk/news/featurednews/title_686716_en.html

A1 – Organising stimuli into zones: Four zones and two ladders

Green zone

This is the zone in which your horse shows relaxed facial and body features in response to the stimulus. His eyes are almond shaped, the edge or corner of his bottom lip is visible, he does not have any wrinkles around his nose and mouth, his nostrils are long, his muscles are soft, and his tail is loose and relaxed.

He does not give any calming signals. If he does, it is only one. If he gives two or three, there is a longer time in between the signals. After the calming signal has been given, the facial features stay relaxed.

Calming signals are: blinking, looking away, half closing the eyes, chewing, tongue-out chewing, yawning, jaw stretch, head turn, neck turn, neck shake, body shake, see-saw lowering, sustained lowering, curving, splitting, showing the hindquarters, showing the flanks, eating, immobility, and slowing down.

'Calming signals are behaviours that are aimed at appeasement, at avoiding conflict, to be polite. Displaying these behaviours allows a horse to get through a situation while keeping social relations intact. Calming signals also enable possible tension to discharge' (Draaisma, 2017)[1].

Yellow zone

In response to the stimulus, the horse has one or two facial features that stay tense for a longer time. These can be round eyes and/or round nostrils. The mouth is still relaxed. The tail is loose or lightly carried.

He gives one or more calming signals at the same time or in quick succession. He may also show minimal displacement behaviour.

Displacement behaviours are: sniffing or stirring up the ground without eating, rubbing the head and neck along his own leg, rubbing the head and neck against objects, self-biting, licking objects, pawing, rolling, and head swing.

'Displacement behaviours are meant to discharge tension. By focusing on something else, you can block out the circumstance around you. Displacement activities involve physical behaviours that fall into a different behavioural group than the behaviours that occur before them.' (Draaisma 2017).

Orange zone

The horse has three facial features that are tense. For example, he may have round eyes, round nostrils, and a tense mouth, with lips that are pressed closed parallel to each other. He can show multiple calming signals and displacement behaviours that alternate. His head is in a mid-high to high position. At times, his muscles are tensed somewhat. His tail is lightly to moderately carried.

He can also have moments in which he withdraws into himself with his head and neck in a half-low to low position, an inward gaze, and a passive posture without movement.

Red zone

The horse has three tense facial features. His eyes and nostrils are round, his mouth and nose have a different shape (an elongated upper lip and a pointed or dented nose), and his chin is pulled up. The muscles in his body are tense; his head and neck position is high; his tail is carried lightly to high; he eats or drinks in a rushed way or stops eating and drinking; he may defecate; he clearly wants to escape the situation/stimuli/stimulus; or he may, in very rare situations, want to chase the stimulus away.

He may also turn inward for a longer time, during which his head and neck position is mid-low to low, his gaze is turned inward, and his posture is passive and without movement.

ADDITIONAL REMARKS

During some period of time, a pendulum-like picture may form, in which the horse alternately shows behaviour from two or even three zones. In that case, you can put the event, situation, or stimulus on the border between two zones, enter it in both zones, or enter it in the higher of the two zones in question.

It could be that your horse is so comfortable in a new situation or around new stimuli that he shows no altered body features or signals at all. However, it can also be that when he first encounters a new stimulus/stimuli, his eyes change shape and his nose becomes rounder as he processes the stimulus/stimuli for a brief moment. You can use this as a baseline for your observations, but you can also start a few seconds after your horse makes the encounter for the first time, noting the result of the process following this initial encounter. Of course, you can also make a note of both.

When using the worksheet (A3), you can use the 'further comments' section to jot down things that might be of influence, such as pain, illness, weather type, changing pasture buddies, etc.

Tip: You should work to improve your observation skills. Try to focus as much as possible on the horse's direct expressions without labelling them right away with overarching terms such as 'insecurity', 'submissiveness', or 'misbehaviour'. Also turn off the sound on the video footage. This is another way to decrease the likelihood that your observation technique will be influenced by 'the voice on the tape' or your own preconceived notions.[2]

Do you want to know more about the body features and communicative signals of horses? Books about the body language of horses can be a useful tool. You can also get the information from my book *Language Signs and Calming Signals of Horses,* which contains extensive explanations and 275 pictures of facial features and signals you can see in horses when they are relaxed or when they are experiencing various levels of tension. Some of these features and signals were placed into a ladder, so that you can see at a glance how much tension (or relaxation) your horse is experiencing in response to a stimulus/stimuli or situation in his environment. You will also find two of those ladders here.

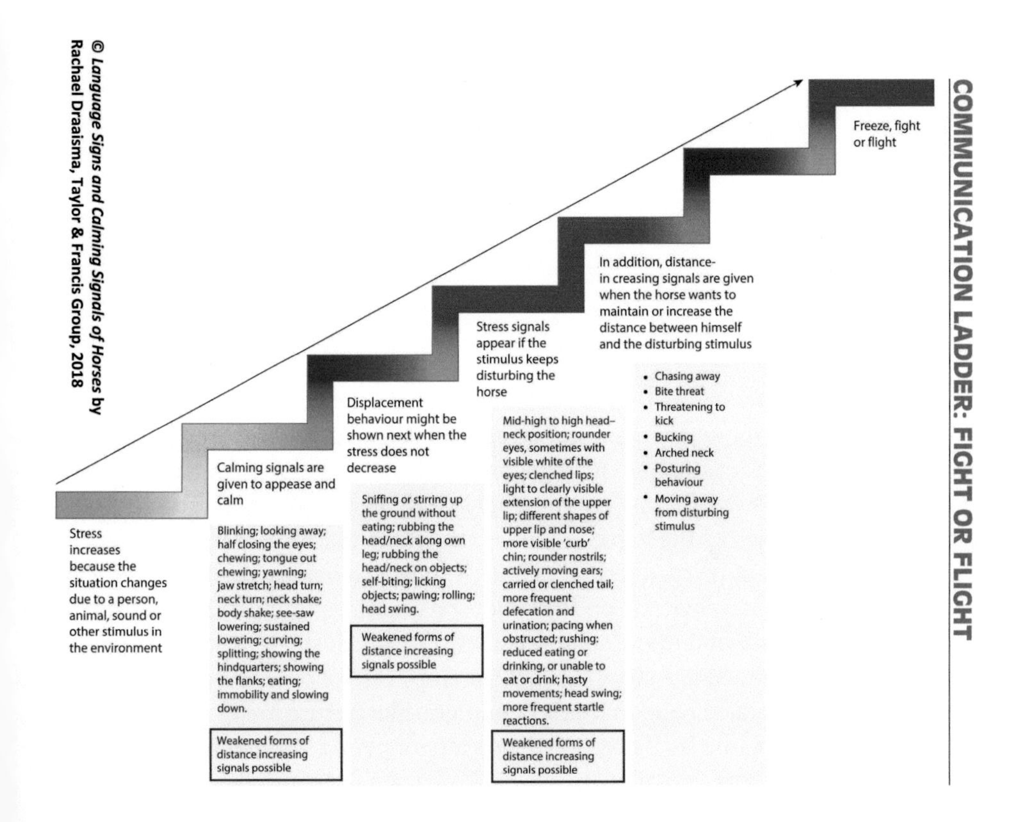

Fig. A1 shows the possible body features and signals of a horse who is dealing with a stimulus/stimuli/situation that is causing rising tension in him. Depending on the situation/stimulus/stimuli and the horse and handler, this build-up of tension can stop and bend downward at any point on the ladder (see Fig. A2). The features and signals your horse displays can be clearly demarcated and neatly follow one another one rung at a time. However, a pendulum-like picture can also form for a time, in which the tension builds and then decreases again. It is also possible that the tension rises so quickly that the displacement behaviours are skipped altogether.

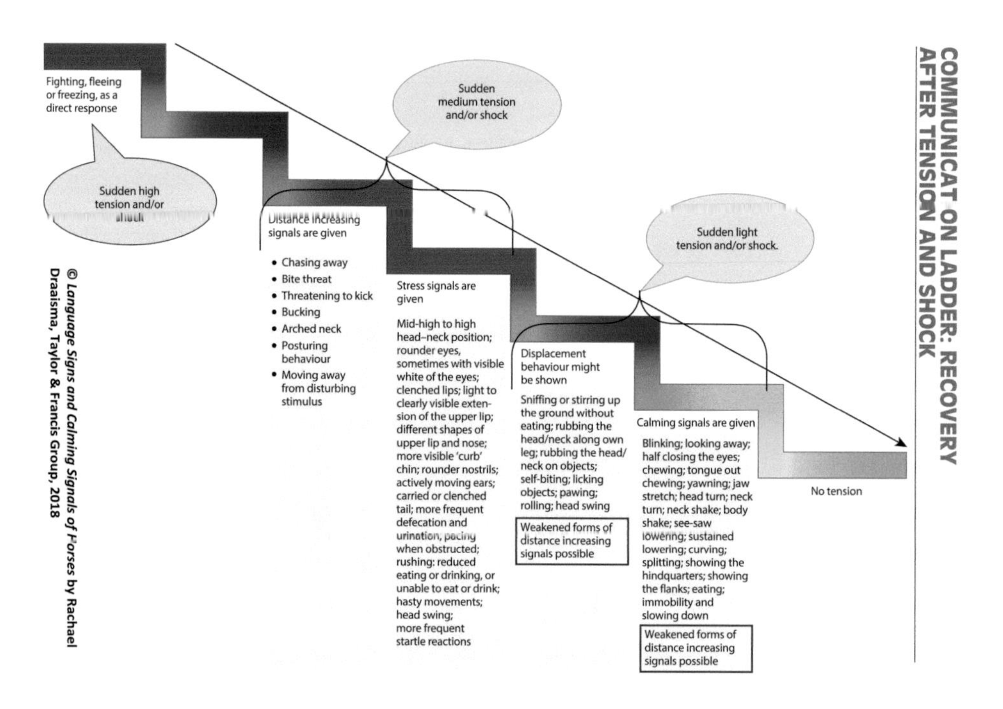

Fig. A2 reflects the situation of a horse who suddenly experiences tension. This can be light tension or a shock, for instance, when a wheelbarrow tips over near him. It can also be medium tension, high tension, or shock, for instance when a motorcycle suddenly revs up close by. Situations that might cause light, medium, or high tension are things such as the horse moving to a different facility, or having to stay overnight at a competition or veterinary clinic. The length of time it takes the horse to become relaxed again depends on the seriousness of the stimulus/situation and the nature of the horse and his handler (the extent to which she is able to change the situation). For this ladder, it is also true that the steps can follow one another clearly and evenly, but it can also happen that a pendulum-like picture arises, or that the horse maintains the same level of tension for a longer period without recovering in any way.

REFERENCES

1. Draaisma R. (2017) *Language signs and calming signals of horses-recognition and application*. CRC Press.
2. Bell C., Rogers S., Taylor J., Busby D. (2019) *Improving the recognition of equine affective states*. https://www.ncbi.nlm.nih.gov/pmc/articles/PMC6941154/

A2 – Zone worksheet

Horse:

Location:

Further comments:

Date:

Start and end time:

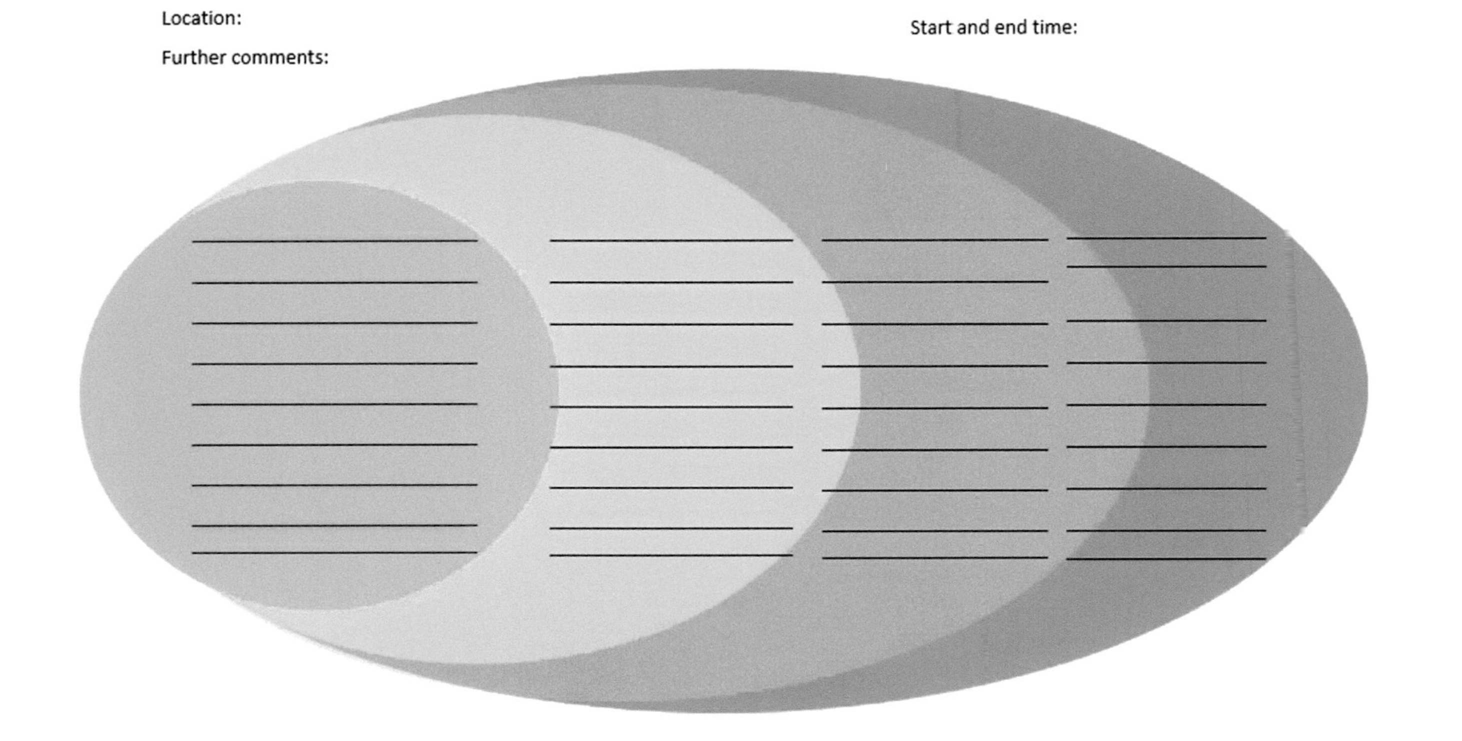

A3 – The haystack exercise

Preparation: Create little piles of hay in several places around the facility. These piles are at easily visible distances from one another.

Phase 1: Lead your horse to the haystacks on a long rope. When your horse sees them and wants to approach them, you follow him. He is allowed to eat all the piles, and he can choose in which order he does this. You stand by your horse when he is eating, and you follow him when he walks to the next haystack. As you walk, you practice taking different positions. For instance, you walk beside your horse's head, his shoulder, his torso, his hindquarters, or just behind him. If you are both comfortable with this and your horse is engaged as he goes from stack to stack, you move to Phase 2.

Phase 2: Introduce hand gestures or a vocal cue. The hand gesture or sound means 'come with me'. Be careful not to use the same sound for this that you also use to spur your horse on.

Again, lay out several haystacks. The moment your horse starts to move to the next haystack, you make the 'come with me' hand gesture or vocal cue (make sure you are standing beside his head or shoulder when you do this), thereby, in a way, making his plan to go to the next haystack yours as well: 'come to this haystack (where you were heading anyway)'. In addition, when your horse stops at a haystack (which he had already intended to do), you can make a hand gesture as a stop signal, possibly with an accompanying word.

Phase 3: Put down several small haystacks, but this time, put small pieces of food your horse likes in between the stacks on the ground, such as a carrot or an apple. The moment the horse starts to walk toward a haystack, you 'discover' the carrot or apple pieces, you move ahead. You bring him along using the hand gesture or vocal cue, and stop him using the stopping sound or hand gesture at the location of the carrot.

That way, you alternate roles: sometimes your horse leads you to food; other times, it is the other way around.

This exercise is simple, but it contains a number of good goals: It improves the relationship between horse and handler (you are working on reciprocity), it helps the horse to learn to understand basic gestures, and it allows you to practice taking different positions while you are walking with your horse (also useful for tracking). In addition, it is a healthy exercise for horses who live with long breaks between meals.

A4 – Checklist 1

For laying a trail when horse and handler are only just starting to track for the scent bag.

- Is your horse accustomed to eating from the bag?
- Is your horse accustomed to the bag being pulled from the ground?
- How did the previous trail go?
- How is the horse feeling today? Does he seem tired, tense, or, to the contrary, physically and mentally fit? (Incorporate this into your plans.)
- What space/location did you pick? One that is as stimulus-free as possible, or one that is more difficult?
- Is the horse comfortable in the space? Adapt the trail accordingly.
- Are there droppings or other things in the space that might distract the horse? Do you want this or not?
- Is there a breeze? If there is, how will you lay the trail? In the direction of the wind or against it? Or does the wind blow sideways across the trail?
- Are you wearing good, old, recognizable shoes?
- How deep an imprint into the ground will you make with these shoes?
- How will you mix things up? Will you create a longer trail, take a different turn, or more than one turn? How sharp do you make these? (Not too sharp in the beginning.)
- Where do you enter and exit the space yourself?
- Where do you put the smeller? Can it blow away in the breeze?
- Does the smeller not smell too strongly of laundry detergent, or does it have a neutral smell, which will allow your horse to really smell the scent of your shoes?
- Have you thoroughly wiped your feet on the smeller?
- Have you put food rewards into the scent bag that will allow your horse to eat for at least 2–3 minutes?
- Does this reward include your horse's favourite treats? Are the foods fresh and moist so that they secrete a lot of scent?
- Before tracking, did you loosen your control of the horse so that he feels free enough to take the initiative?
- Walk slowly toward the smeller (in order to break the familiar marching pattern).
- Does your horse feel comfortable enough around you to take charge and take turns to the left and the right, and is he unafraid to take the food?
- Are your facial and body features relaxed? Generally, do not give verbal or nonverbal clues while your horse is tracking.
- Does your horse want to eat while tracking? Give him this break; he will continue with the trail eventually. If he does not, evaluate and make a new plan.
- Did you put out food and water so that the horse can eat and drink before and after tracking?

A5 – Checklist 2

For the horse and handler who are just starting to track for the scent bag, but who have not yet been successful in finding the bag, these points hopefully will help you to make improvements for next time. Good luck!

- Was the ground unspoiled before you laid the trail, or did you accidentally leave other trails at an earlier time?
- Is the smeller a different piece of cloth than usual? Is it fairly scent neutral, apart from your footsteps? (In other words, does it not smell strongly of detergent?)
- What shoes are you wearing? The same ones as last time? Are these your most suitable shoes, ones you have worn often and that have deep grooves?
- What is the direction of the wind?
- Has the wind perhaps scattered the trail you laid?
- Are there droppings or other things in the arena that might have distracted your horse?
- Does your horse feel comfortable enough in the location?
- Does your horse feel comfortable enough to be here with you (without other horses)?
- How did you assist your horse? Are improvements possible?
- Does your horse feel comfortable enough to take the lead, walk in front, and turn to the left or right?
- Did you miss your horse pointing at the bag?
- Is your horse not afraid to point to the scent bag?
- Did you use your horse's favourite food rewards? Were they moist enough and not dried out?
- Has your horse learned that the smeller is the starting point for finding the food?
- Is your horse feeling okay today – not too tired or so tense that he can no longer track?
- Is he in pain?
- Has he been able to peek at where you buried the bag?
- Is he distracted by other scents, sounds, or activities around him?
- Is the weather a bit too difficult for him? (Warm, dry air, or pouring rain make things harder.)
- Has he become so excited during the run-up to tracking that he is now no longer able to calm himself down?
- Is the horse too hungry, making him too eager to find the food to calm himself down enough to track? (In that case, let your horse eat something first.)
- Is the horse too excited from experiences he has had in the past few days to hours? (See if he can calm himself down during tracking, and otherwise let him graze and try again another time).

- Is the trail you laid too difficult because you covered too great a distance too quickly? Is the trail too easy?
- Did you lay a trail that is comparable to the one you laid the last time(s), so that the horse skips the scenting, but works on his well-developed cognitive map and walks directly toward the spot where he thinks the scentbag is hidden?

Bibliography

BOOKS

Bekoff M. (2008) *The emotional lives of animals.* New World Library, Novato.

Bekoff M., Pierce J. (2019) *Unleashing your dog – A field guide to giving you canine companion the best life possible.* New World Library.

Blake A. (2019) *Going steady: More relationship advice from your horse.* Prairie Moon Press.

DeGiorgio F., Schoorl J. (2014) *The cognitive horse: An inspiring journey towards a new co-existence.* Learning Animals, Nistelrode.

Gonzalez J. (2018) *Equine empowerment-a guide to positive reinforcement training.*

Horowitz A. (2016) *Being a dog-following the dog into a world of smell.* Scribner, New York.

Horowitz A. (2013) *Met Andere Ogen-Manieren om meer waar te nemen.* Uitgeverij Balans.

Ingraham C. (2014) *How animals heal themselves.* Ingraham Trading Limited.

Kainer R.A., McCracken T.O. (1998) *Horse anatomy: a coloring atlas,* Second Edition. Alpine Puclications, Crawford.

Knaapen R. (2012) *Coachen met paarden – Het systemisch perspectief.* Boom uitgevers Amsterdam.

Leibbrandt K. (2018) *Je paard succesvol trainen-Paardvriendelijk presteren met het moderne sportpaard.* Maarten Beernink, Het Boekenschap.

McBane S. (2012) *Horse senses.* Manson Publishing Ltd.

McDonnell S. (2003) *The equid ethogram: A practical field guide to horse bahaviour.* Blood Horse Inc.

Neugebauer G.M., Neugebauer J.K. (2011) *Het gedrag van paarden beter begrijpen.* Eugen Ulmer KG, Stuttgart, Germany.

Robertson J. (2010) *The complete dog massage manual.* Hubble & Hattie, Poundbury.

Robertson J., Mead A. (2013) *Physical therapy and massage of the dog.* Manson Publishing, London.

Rugaas T. (2002) *Kalmerende Signalen, wat u en uw hond elkaar vertellen.* Yggdrasil.

Rugaas T. (2008) *Blafgedrag van honden, watbetekent het en hoe kun je ermee omgaan.* Yggdrasil.

Swaab D. (2010) *Wij zijn ons brein, van baarmoeder tot alzheimer.* Olympus.

Teblick I. (2012) *Denkwerk of waarom paarden van leertheorie houden.* Crème Brûlée, Bavarois bvba.

VanFleet R. (2013) *The human half of dog training-collaborating with clients to get results.* Dogwise Publishing.

VanFleet R., Faa-ThompsonT. (2017) *Animal assisted play therapy.* Professional Resources Press, Sarasota.

Waal de F. (2016) *Zijn we slim genoeg om te weten hoe slim dieren zijn? (Are we smart enough to know how smart animals are?).* Atlas Contact.
Weston H., Bedingfield R. (2019) *Connection training: The heart and science of positive horse training.* Connection Training.

STUDIES

Equimed (2010) *Recognizing and preventing equine stereotypies.* https://equimed.com/health-centers/behavior/articles/recognizing-and-preventing-equine-stereotypies
Fureix C., et al. (2012) *Towards an ethological animal model of depression? A study on horses.* https://journals.plos.org/plosone/article?id=10.1371/journal.pone.0039280
Haesler S. (2020) *Novelty speeds up learning thanks to dopamine activation.* http://www.vib.be/en/news/Pages/Novelty-speeds-up learning-thanks-to-dopamine-activation.aspx
Henry S., et al. (2009) *Neonatal handling affects durably bonding and social development.* https://journals.plos.org/plosone/article?id=10.1371/journal.pone.0005216
Ijichi C., et al. (2018) *Dually noted: the effects of a pressure headcollar on compliance, discomfort and stress in horses, during handling.* https://www.researchgate.net/publication/323120474_Dually_Noted_The_effects_of_a_pressure_headcollar_on_compliance_discomfort_and_stress_in_horses_during_handling
Kieson E., Abramson C. (2016) *Equines as tools vs partners: A critical look at the uses and beliefs surrounding horses in equine therapies and argument for mechanical horses.* https://www.researchgate.net/publication/309705463_Equines_as_tools_vs_partners_a_critical_look_at_the_uses_and_beliefs_surrounding_horses_in_equine_therapies_and_argument_for_mechanical_horses
Kloet de E. (2009) *Stress: Neurobiologisch perspectief.* http://www.tijdschriftvoorpsychiatrie.nl/assets/articles/articles_2814pdf.pdf
Mariti C., et al. (2010) Domestic dogs display calming signals more frequently towards unfamiliair rather than familiair dogs. *J Vet Behav* 5(1):62–63.
Mariti C., et al. (2014) Analysis of calming signals in domestic dogs: Are they signals and are they calming? *J Vet Behav* 9(6).
McEwen, B.S. (2008) Central effects of stress hormones in health and disease: Understanding the protective and damaging effects of stress and stress mediators. *Eur J Pharmacol* 583(2–3):174–185.
Mediavilla C., et al. (2016) *Role of anterior piriform cortex in the acquisition of conditioned flavour preference.* https://www.ncbi.nlm.nih.gov/pmc/articles/PMC5022059/
Moulton D.G. (1967) *Olfaction in mammals.* https://academic.oup.com/icb/article/7/3/421/244992

Neut van der D. (2007) *Kunnen dieren denken*. http://www.dagmarvanderneut.nl/wetenschapsjournalistiek_files/psy04_dierenbrein.pdf

Perry P., et al. (2020) *A comparison of four environmental enrichments on adoptability of shelter dogs*. https://www.sciencedirect.com/science/article/pii/S1558787819300292

Ruet A., et al. (2019) *Housing horses in individual boxes is a challenge with regard to welfare*. https://www.mdpi.com/2076-2615/9/9/621

Schrimpf A., et al. (2020) *Social referencing in the domestic horse*. https://www.researchgate.net/publication/338674058_Social_Referencing_in_the_Domestic_Horse

Scopa C., et al. (2018) *Physiological outcomes of calming behaviors support the resilience hypothesis in horses*. https://www.ncbi.nlm.nih.gov/pmc/articles/PMC6269543/

Seligman, M.E., & Maier, S.F. (1967) Failure to escape traumatic shock. *J Exp Psychol* 74(1):1–9.

SOME STUDIES ON THE OLFACTORY SYSTEM OF HORSES

Baldwin A.L., Shea I. (2018) *Effect of aromatherapy on equine heart rate variability.* https://www.sciencedirect.com/science/article/abs/pii/S0737080618301266

Berg van den M., Giagos V. (2016) *The influence of odour, taste and nutrients on feeding behaviour and food preferences in horses*. https://www.sciencedirect.com/science/article/abs/pii/S0168159116302477

Berger J.M., et al. (2013) *Behavioral and physiological responses of weaned foals treated with equine appeasing pheromone: A double-blinded, placebo-controlled, randomized trial*. https://www.sciencedirect.com/science/article/abs/pii/S1558787812001219

Boniface M.K., Hassanali J. (2011) *Comparative morphometry of the olfactory bulb, tract and stria in the human, dog and goat*. https://www.researchgate.net/publication/261416780_Comparative_Morphometry_of_the_Olfactory_Bulb_Tract_and_Stria_in_the_Human_Dog_and_Goat

Christensen R. (2008) *Predator odour per se does not frighten domestic horses*. https://www.researchgate.net/publication/248336335_Predator_odour_per_se_does_not_frighten_domestic_horses

Cozzi A., Lafont Lecuelle C. (2013) *The impact of maternal equine appeasing pheromone on cardiac parameters during a cognitive test in saddle horses after transport*. https://www.researchgate.net/publication/257626438_The_impact_of_maternal_equine_appeasing_pheromone_on_cardiac_parameters_during_a_cognitive_test_in_saddle_horses_after_transport

Crowell-Davis H. (1985) *The ontogeny of flehmen in horses*. https://www.sciencedirect.com/science/article/abs/pii/S0003347285800051

Deshpande K., et al. (2018) *The equine volatilome: Volatile organic compounds as discriminatory markers*. https://www.sciencedirect.com/science/article/abs/pii/S0737080617304318

Ferguson C., Kleinman H.F. (2013) *effect of lavender aromatherapy on acute-stressed horses*. https://www.sciencedirect.com/science/article/abs/pii/S0737080612002183
Jezierski B., et al. (2015) *Excreta-mediated olfactory communication in konik stallions: A preliminary study*. https://www.researchgate.net/publication/275219625_Excreta-mediated_olfactory_communication_in_Konik_stallions_A_preliminary_study
Jezierski B., et al. (2018) *Do olfactory behaviour and marking responses of konik polski stallions to faeces from conspecifics of either sex differ?* https://www.ncbi.nlm.nih.gov/pubmed/28962880
King, G. (2006) *Scent-marking behaviour by stallions: An assessment of function in a reintroduced population of Przewalski horses* (Equus ferus przewalskii). https://www.researchgate.net/publication/230013809_Scent-marking_behaviour_by_stallions_An_assessment_of_function_in_a_reintroduced_population_of_Przewalski_horses_Equus_ferus_przewalskii
Krueger F. (2010) *Olfactory recognition of individual competitors by means of faeces in horse* (Equus caballus). https://www.researchgate.net/publication/49659137_Olfactory_recognition_of_individual_competitors_by_means_of_faeces_in_horse_Equus_caballus
Kupke A., et al. (2016) *Intranasal location and immunohistochemical characterization of the equine olfactory epithelium*. https://www.ncbi.nlm.nih.gov/pmc/articles/PMC5061740/
Lampe A. (2012) *Cross-Modal recognition of human individuals in domestic horses* (Equus caballus). https://link.springer.com/article/10.1007/s10071-012-0490-1
Lee K., et al (2016) *Histological and lectin histochemical studies of the vomeronasal organ of horses*. https://www.sciencedirect.com/science/article/abs/pii/S0040816616300076
Micera E., et al (2012) *Reduction of the olfactory cognitive ability in horses during preslaughter: Stress-related hormones evaluation*. https://www.sciencedirect.com/science/article/abs/pii/S0309174011002324
Peron F. (2015) *Horses* (Equus caballus) *discriminate body odour cues from conspecifics*. https://www.researchgate.net/publication/259201807_Horses_Equus_caballus_discriminate_body_odour_cues_from_conspecifics
Rubenstein H. (1992) *Horse signals: The sounds and scents of fury*. https://www.researchgate.net/publication/226982775_Horse_signals_The_sounds_and_scents_of_fry
Schmidt M., Knemeyer C. (2019) Neuroanatomy of the equine brain as revealed by high-field (3Tesla) magnetic-resonance-imaging. *PLoS One* 14(4):e0213814, https://www.ncbi.nlm.nih.gov/pmc/articles/PMC6443180/

Siniscalchi M., et al. (2015) *Right-nostril use during sniffing at arousing stimuli produces higher cardiac activity in jumper horses.* https://www.researchgate.net/publication/271591811_Right-nostril_use_during_sniffing_at_arousing_stimuli_produces_higher_cardiac_activity_in_jumper_horses

VanSommeren A., VanDierendonck M. (2010) *The use of equine appeasing pheromone to reduce ethological and physiological stress symptoms in horses.* https://www.sciencedirect.com/science/article/abs/pii/S155878780900327X?via%3Dihub

Index

D

E

F

L

M

N

O

P

U

V

W

Y

Z

For Product Safety Concerns and Information please contact our
EU representative GPSR@taylorandfrancis.com Taylor & Francis
Verlag GmbH, Kaufingerstraße 24, 80331 München, Germany